MATHEMATICAL
PLATO

Roger Sworder

MATHEMATICAL
PLATO

ANGELICO PRESS
SOPHIA PERENNIS

First published in the USA
by Angelico Press / Sophia Perennis
© Roger Sworder 2013

Series editor: James R. Wetmore

For information, address:
Angelico Press, 4619 Slayden Rd. NE
Tacoma, WA 98422
www.angelicopress.com

Library of Congress Cataloging-in-Publication Data

Sworder, Roger.
Mathematical Plato / Roger Sworder.—First ed.

pages cm.

ISBN 978-1-59731-138-0 (pbk: alk. paper)
1. Plato. 2. Science—Philosophy.
3. Mathematics—Philosophy. 4. Ancient philosophy.
I. Title
B 395.88 2013
184—dc22   20130115688

Cover Design: Michael Schrauzer

CONTENTS

Acknowledgments

I AM grateful to Professor Terry Mills for correcting some of the mathematics in my manuscript, and to Dr. Dorothy Avery for preparing it for publication.

Chapters 6, 7, and 9 appeared in earlier forms in *A Contrary History of the West*, and chapter 7 in *Parmenides at Delphi*.

Preface

THE TITLE of this book is surely as unattractive at this time as a title can be. Plato himself has been at odds with his own institution since Francis Bacon. In our time he is the very paradigm of classist, sexist, and racist patriarchy. In those places where his books are still taught, there will always be students who sing to his poetry, but these are just the students who will be deterred by the other word of my title. Among Humanities students the distaste for mathematics is very widespread.

And yet according to its title at least, this book should be an Academic book of books, the quintessence of Western learning and the great goal of our enterprise. Plato founded the Academy and placed mathematics at the center of its work. He established as no one else the unbreakable connection between the natural and the mathematical sciences. This connection has never been stronger than it is now. The mathematical passages in his Dialogues are often abstruse, and these passages are arguably the real core of his teaching, difficult precisely so as to draw out the full powers of the student.

There is one qualification which any reader of this book must meet. They must have copies of the *Phaedo*, the *Republic*, and the *Timaeus*. Very little of these texts appears in this book, but most of this book is a detailed and intricate study of Plato's actual words. It is much easier to follow arguments about texts if the text itself is present to the eye simultaneously, and the easiest way to do this is to have both books open at the right pages. Any translation will serve, provided only that the reader has most confidence, for the moment, in the brief translations here of key passages.

For The School of Philosophy

Melbourne, Australia

PART I

MATHEMATICS and NATURAL SCIENCE

1

Plato's Grand Theory

IN HIS DIALOGUE the *Parmenides* Plato makes an extended attack on his own theory of ideas or forms. This theory is the basis of his psychology in the *Phaedo*, of his ideal state in the *Republic* and of the natural creation in his *Timaeus*. But his attack on this theory goes undefended and unanswered either in the *Parmenides* or elsewhere. The dating of these various dialogues is uncertain and so therefore are their temporal relations to each other. If the *Parmenides* was written after the others, then maybe Plato intended it to supersede them. In that case Plato's intellectual career may be compared to Wittgenstein's, who rejected the theory of language expounded in his early *Tractatus* for the different approach of the *Philosophical Investigations*. And from Gilbert Ryle onwards a number of Platonic scholars have supposed exactly this, that Plato changed his mind and became more of a linguistic analyst like themselves in his later years.

But even it if could be shown that the *Parmenides* was composed later than the works which depended on the theory of ideas, this would make no difference to the chronology or sequence in which Plato himself presents them. And in that sequence the *Parmenides* is clearly first, not last. The *Phaedo* presents Socrates at his death, the *Republic* and the *Timaeus* show him as mature, the *Parmenides* as very young indeed for serious discussion. But if the Socrates of the *Phaedo, Republic* and *Timaeus* still believes in the theory of ideas after the trouncing the young Socrates was given in the *Parmenides,* then Plato's Socrates at least had no change of heart about the theory at depth, but continued to propound it to the end. That Plato did the same and held to the theory of ideas throughout his life was the unanimous view of antiquity.

The *Parmenides* presents a great many arguments against the theory of ideas but there is one which has always attracted inordinate attention. I present it here in Cornford's translation. Parmenides is the questioner, Socrates the respondent:

> How do you feel about this? I imagine your ground for believing in a single form in each case is this. When it seems to you that a number of things are large, there seems, I suppose, to be a certain single character which is the same when you look at them all, hence you think that largeness is a single thing.

> True, he replied.

> But now take largeness itself and the other things which are large. Suppose you look at all these in the same way in your mind's eye, will not yet another unity make its appearance — a largeness by virtue of which they all appear large?

> So it would seem.

> If so, a second form of largeness will present itself, over and above largeness itself and the things that share in it, and again, covering all these, yet another, which will make all of them large. So each of your forms will no longer be one, but an indefinite number.[1]

And

> Well, if a thing is made in the image of the form, can that form fail to be like the image of it, in so far as the image was made in its likeness? If a thing is like, must it not be like something that is like it?

> It must.

> And must not the thing which is like share with the thing that is like it in one and the same thing [character]?

> Yes.

> And will not that in which the like things share, so as to be alike, be just the form itself that you spoke of?

1. *Parmenides* 132 a.1–b.3.

2

Certainly.

> If so, nothing can be like the form, nor can the form be like any-
> thing. Otherwise a second form will always make its appearance
> over and above the first form, and if that second form is like any-
> thing, yet a third. And there will be no end to this emergence of
> fresh forms, if the form is to be like the thing that partakes of it.[2]

Aristotle recast this argument in terms of the form of Man or idea
of Man and individual men, but the regress was the same, an infi-
nite series of ideas of Man. So this argument against Plato's theory
of ideas is known as the Third Man argument. But the argument is
of the most particular concern to contemporary philosophers
because it exploits a very real weakness in the theory of ideas as they
understand it. From their point of view the theory of ideas con-
founds two very different things: the notion or concept of something with the characteristic property or quality of that something.
So, to put it at its most absurd, the notion of proximity must itself
be nearby, and the notion of circularity must itself be round. The
Third Man argument, like the argument about largeness, shows that
such notions cannot themselves share in the properties or attributes
which they conceive or name. There is a great divide between our
thoughts and language on the one side and the actual world on the
other. The theory of ideas had fundamentally misrepresented this,
and the Third Man argument was Plato's acknowledgment of his
mistake. Furthermore this mistake is incorrigible and fatal. And so
the later Plato becomes one of us, a visionary chastened by the
actual facts of language and thought.

Here, for example, is Gilbert Ryle on Plato's theory of ideas or
forms in his essay on Plato's *Parmenides,* published in 1939:

> We object at once that of course concepts like magnitude, equality,
> smallness and the rest do not themselves have magnitudes. Bigness
> is not bigger or smaller than anything else, nor equal in size
> to anything else. It is nonsense to ascribe predicates of size to
> concepts of size. Attributes such as quantitative dimensions, are
> not instances of themselves. Indeed, like Professor Taylor or Mr.

---

2. *Parmenides,* 132 d.3–133 a.2

Hardie, we are ready to declare with confidence that no 'universal', i.e. no quality, relation, magnitude, state, etc., can be one of its own instances. Circularity is not circular nor is proximity adjacent. Nor even are such concepts capable of being instances of other concepts of the same family as themselves. It is nonsense to describe circularity as circular *or* of any other shape; and it is nonsense to describe redness as of *any* color, equality as of *any* dimensions. We are right to make such objections. The theory of Forms is logically vicious if it implies that all or some universals are instances of themselves or of other universals of the same family with themselves. And Plato had, apparently, once thought that beauty was beautiful and goodness was good; maybe he had thought that circularity and circularity alone was perfectly circular.

But that such descriptions of qualities, magnitudes, relations, etc., are illegitimate has to be shown and not merely felt. Plato is either showing it or on the way to showing it in this part of the dialogue.[3]

For Ryle the theory of ideas or forms is 'logically vicious' and Plato was either aware of this or very nearly so in the Third Man argument of the *Parmenides*. And, of course, from Ryle's point of view, the theory really is logically vicious because it makes absurd mistakes about concepts as we understand them. But that presupposes that Plato really wanted to think and talk about what we think of as concepts, and this presumption is premature.

In 1954 Gregory Vlastos published an analysis of the Third Man argument in the *Parmenides*. This essay claimed to expose a series of logical and not merely conceptual errors in Plato's formulation of his theory of ideas. Vlastos agreed with Ryle that in the *Parmenides* Plato had begun to realize that his wonderful theory had weaknesses. But Vlastos expands on Plato's psychological and philosophical situation:

This is Plato's mood in the *Phaedo*, the *Symposium*, and the *Republic*. The Theory of Forms is then the greatest of certainties, a place of unshakable security to which he may retreat when doubt-

3. *Studies in Plato's Metaphysics*, ed. R. E. Allen (New York, Routledge & Kegan Paul, 1967) p. 103.

ful or perplexed about anything else. But as he lives with his new theory and puts it to work, its limitations begin to close in upon him. He begins to feel that something is wrong, or at least not quite right, about his theory, and he is puzzled and anxious. If he has courage enough, he will not try to get rid of his anxiety by suppressing it. He may then make repeated attempts to get at the source of the trouble, and if he cannot get at it directly he may fall back on the device of putting the troublesome symptoms into the form of objections. He can hardly make these objections perfectly precise and consistent counter-arguments to his theory. Unless he discovers the exact source of its difficulties and can embody the discovery, the objections are likely to be as inadequate in their own way as is their target. They will be the expression of his acknowledged but unresolved puzzlement, brave efforts to impersonate and cope with an antagonist who can neither be justly represented nor decisively defeated because he remains unidentified and unseen. This, I believe, is an exact diagnosis of Plato's mind at the time he wrote the *Parmenides*.[4]

Like Ryle, Vlastos supposes that Plato is on his way to an adequate conception of concepts, which is to say a conception like Vlastos' own. Vlastos emphasizes Plato's 'honest perplexity' over the issue and his refusal to avoid the problems which his theory raised. They were real and unanswerable problems and with them the theory which underpinned his greatest works to that point had collapsed.

According to Vlastos, Plato was misled by certain features of the Classical Greek language:

In this case there is a further factor, and a very prosaic one, which may blinker the logical vision of a clearheaded man. It is the fact that 'Justice is just', which can also be said in Greek as, 'the just is just', can be so easily mistaken for a tautology, and its denial for a self-contradiction. I am not suggesting that the Assumption of Self-Predication is just a symptom of the tyranny of language over ontology. The suggestion would not even be plausible, for other philosophers, using the same language, made no such assumption. The assumption has far deeper roots, notably religious ones, which I cannot explore in this paper. What can be debited to lan-

4. *Studies in Plato's Metaphysics*, op.cit., p. 255.

guage is simply the fact that an assertion which looks like an identity-statement may be taken as having the certainty of a tautology; and the illusion of its self-evidence could very well block that further scrutiny which would reveal that it implies a proposition which so far from being self-evident leads to self-contradiction.[5]

On such grounds, then, Ryle and Vlastos and many others concluded that the Third Man argument was valid and the theory of forms or ideas incoherent. With Plato himself as they thought, they dismissed what Vlastos called 'one of the richest and boldest metaphysical theories ever invented in Western thought.' Faint praise, indeed, when its author has forgotten that the metaphysical is never invented!

Vlastos' essay was as productive of reactions as Ryle's had been and among them was an essay by P. T. Geach. In strong language for an academic Geach wrote:

> It would be a crude mistake to regard Plato's Theory as just a 'disease of language' . . . .

Geach had negative and positive points to make. He was very concerned by Vlastos' use of abstract English nouns like 'largeness' and 'F-ness' as designations for Plato's ideas. Such terms came, Geach believed, with a host of tacit assumptions quite alien to Plato's mind, and in no case can we assume that they are what Plato intended either clearly or obscurely. Geach himself has an entirely different view of the Platonic ideas to offer, which he acknowledged that he owed to Wittgenstein:

> The bed in my bedroom is to the Bed, not as a thing to an attribute or characteristic, but rather as a pound weight or yard measure in a shop to the standard pound or yard. (I owe this insight to discussion with Wittgenstein.) This comparison brings in Self-Predication in a way, because we use the same word 'yard' or 'pound' both of the shop-man's weight or measure and of the standard. The explicit Self-Predications 'the standard yard is a yard long', or 'the standard pound weighs a pound', would indeed not ordinarily be made; indeed, we should have some hesitation what to make of

5. Ibid, p. 250.

such statements. Are we to regard them as trivial tautologies? There is, on the contrary, some ground for regarding them as plain absurdities; the one thing that you cannot measure or weigh against a standard is—the standard itself.[6]

To think of a Platonic form or idea as a standard like the Imperial Standard Pound was certainly within the range of Plato's own experience. Since the seventh century the weights and measures of King Phaidon of Argos had served larger and larger areas of Greece, to be modified by Solon in Athens' case. All the more surprising then that Plato did not use this as an example if this is what his ideas were intended to do. In an edgy reply to Geach, Vlastos acknowledged the worth of thinking of the ideas as patterns in this much fuller sense, and claimed that Plato's difficulties with his theory arose from his trying to combine two different and incompatible functions in one: the notion of an attribute or characteristic and the notion of just such a standard. Still, for this student at least, Geach and Wittgenstein helped to break the impasse into which Plato's *Parmenides* and its commentators had forced me.

Geach supposes that Plato's distinction between the idea and an instance of it was taken up into Christian theology in the theory of God. For God is good and wise and some people are good and wise, but personal wisdom and goodness are different from God's, not in degree but in kind. The statement 'God is good' is not the same as 'that man is good' since the first of these statements is close to a tautology. Compared to God people are good only by analogy. Or the other way round: the Imperial Standard Pound is itself a pound in a way but only by analogy. Since then the idea of x is x in a different way from the way in which any particular x is x, the Third Man argument against the theory of ideas is invalid and Socrates is saved.

Thanks to Wittgenstein, Geach's defence of the ideas is imaginative but it is not as effective nor as simple as the defence mounted by one of Vlastos' own colleagues, Harold Cherniss. Everywhere in Plato the ideas are related to their instances as originals to their images. An original and its image are alike but the relation is not

---

6. *Studies in Plato's Metaphysics*, op. cit., p. 267.

symmetrical. Nor does the relation require any third term to explain how an original comes to have the same property as its image. If the Third Man argument is valid, then in every case where there is an original and an image, a regress must occur. It does not, and the same is true of the relation between an idea and an instance of it. In Plato's language, according to Cherniss, the idea is x while an instance of the idea merely has the property x. This is the formula which invalidates the Third Man argument.

Vlastos, according to Cherniss

> explicitly interprets the formula to mean that the ideas are *themselves* attributes or properties of particulars; but this is a complete misapprehension. As the passage in the *Republic* shows, it means that of any character or property, x, that a particular *has*, the *reality* is ὁ ἔστιν x [what is x] which it could not be if it were *had* by anything and which therefore must be independent or 'separate' from all manifestations of itself as a property.[7]

We may add Cherniss's 'complete misapprehension' to Geach's 'crude mistake' as a sign of frustration on the part of the friends of ideas. But this exasperation was not all on one side. Here is W. G. Runciman writing two years after Cherniss:

> It is accordingly distressing to find that Professor Cherniss still retains his conviction that Parmenides' arguments are invalid and were seen by Plato to be so.[8]

The relation between a Platonic idea and an instance of it is the relation between an original and an image of it. One Greek word for 'original' here is *paradeigma*. Clearly then the Platonic idea is not a concept nor yet an attribute or property, since the relations between these and their instances are not the same as the relation between an original and its image. But, again, the relation between original and image is not the same as the relation between standard and instance

7. Ibid., p. 373.
8. Ibid., p. 158.

either, between the Imperial Standard Pound and a poundweight. One might be hard put to distinguish between the Imperial Standard Pound and a poundweight but this cannot arise in the case of original and image since to see that something is an image is necessarily to see that it differs from what it images, however alike in other respects. Otherwise it would not be an image but a replica.

In the *Phaedo* Socrates argues that anyone who knows Simmias and recognizes a picture of him must see that the picture 'falls short' of Simmias himself. In the *Republic* Socrates considers the case where someone sees a picture of a carpenter so lifelike, they mistake it for an actual carpenter. An image may be indistinguishable from its original from a certain point of view, namely that point of view from which it is like its original. This is not how it falls short. This in turn entails that the Platonic ideas are not perfect idealizations of imperfect creatures, of flawed instances of themselves in the world. The ideas do not so transcend the physical world that everything here is irremediably second rate. The ideas are as realized as they may be in the limited perfections all around us.

This goes against our sense of the Socrates who tells us that everything in this world is rolling around between opposite states, and is never really one thing nor the other. But this Socrates is making a different point: the difference between the partiality and mutability of things in this world and the self-same identity of the ideas. We may well be deceived by the picture of the carpenter but, look, it is only a picture. What appeared large now appears small; what is equal to one thing is unequal to everything else; every one which is one here also has parts. But what appeared large may have been truly large: snow, as Aristotle said in this context, is nonetheless white for being white only a day. Equals here may be unequal to other things but they may be as equal to each other as any measurement of ours can discern. What is one here may also be partial but it may be one in ways we had never imagined possible before. But if the instances of an idea are, in their limited ways, as perfect as the idea, why not give away the instances altogether and just have the ideas in all the variety of their limited perfections?

And this, I take it, is Parmenides' strategy as he employs the tactic of the Third Man argument in Plato's dialogue. He wants to reunify ideas with their instances as any Monist would. There can be no doubt that Plato's Parmenides believes that there are unchanging ideas. He says that there could be no discourse without them. So this reunification of ideas with their instances consists in absorbing the instances into the ideas, doing away with the instances. And Parmenides' ideas, like Plato's, are up to this task. Parmenides has no problem with ideas which are and have their own characters or properties. His own dialectical display is full of this later on. What Parmenides objects to in Socrates' theory is the sundering of this world from the ideas, the debasement of our realm:

Neither is anything less but everything is full of being.[9]

So there is an irony to how the enemies of ideas, Ryle, Vlastos, Runciman, have used Parmenides' Third Man argument. What Plato wanted to show was how vulnerable his theory was to the strict monistic idealism of the Eleatics. But the analytic philosophers of the mid-twentieth century have turned these arguments to exactly the opposite purpose, to show that the ideas of Plato and Parmenides are 'logically vicious' as analyses of concepts and that Plato discarded them. For the logical analysts, all that remains after Plato's *Parmenides* are the instances and a 'weak' descriptive pattern of concepts. For young Socrates what remained after the *Parmenides* was a theory of ideas at once more receptive to the light of the One and much more transparently apprehensible in the created order.

Plato's theory of ideas expresses that view of the world around us where everything is very beautiful. The myths of the *Phaedo* and *Republic*, *Phaedrus* and *Statesman* spring from this same visionary power. This is the world of Parmenides' chariot ride to the palace of a Goddess who reveals all things to him; the world of the *Iliad* and the *Odyssey*. Divine and mortal freely mix. Everything is well made and performs at the limit of its potential; our everyday activities are archetypal. It is perhaps a greater injustice than Socrates suffered that Plato should ever be considered a Utopian idealist who despised the world.

We may mistake a picture of a carpenter for a carpenter when the

9. *Parmenides*, 8, 24.

picture appears to us exactly as the actual carpenter would. Similarly we may mistake a physical object for an idea when we see a beloved person as the very idea of beauty itself. This is a very common delusion and no one, I dare say, was more prone to it than Plato, the greatest philosopher of erotic love in our era. Is not this what falling in love is, making this mistake? And is the lover wrong? Is not the beloved a single straight ray of beauty pure? Socrates may talk of mortal trash in his ladder of beauties but he has yet to grow out of the thrill when he sees Charmides' breast beneath his tunic!

Fully to understand how this world is the most perfect possible realization of the fullest totality of the most exquisite ideas is a Herculean education. The deepest seclusion is needed to complete a thorough study of arithmetic, geometry, astronomy, music and dialectic. After these come their applications to the natural sciences. These studies are typically pursued through early adulthood to middle age. That done, the meaning of our human life emerges as a vision in which no further parting is possible between the absolute and the relative, the eternal and the temporal. There is *apocatastasis*. This is the goal of Plato's theory.

2

Recollection

Phaedo 73b–77a

TO UNDERSTAND Plato's mathematics we must understand his theory of mathematical ideas. Strikingly, Plato derived our knowledge of mathematical ideas from prenatal memory and not only from our experiences in this world. Plato explains how our prenatal memories apply mathematical ideas to the world of our senses by his theory of recollection. The most extended account of recollection is in the *Phaedo* and is the subject of this chapter. The account of recollection in the *Phaedo* underpins the fuller epistemology of mathematics in the *Republic*.

The general points made in the preliminary exposition of Socrates' theory of recollection from Phaedo 73c.1 to 74c.8 are here numbered a1, a2, a3, etc. The parallel points, the application of the general to the case of equal things and the Equals themselves, are here numbered b1, b2, b3, etc.

> a1) 'If someone is to be reminded of something, he must have known it at some time before' (73c.1).

> a2) 'If someone who has seen or heard or in some other way perceived something knows not only that something but also thinks of another thing, the knowledge of which is not the same but different, he is reminded of that of which he thinks' (73c.6–10).

Here Socrates gives an example, the example of the lover being reminded of his beloved by his beloved's lyre (73d.5–9). The satisfaction of this condition alone establishes a case as one of being reminded (73d.8 'This is being reminded').

a3) 'The most extreme case of being reminded is being reminded of things which one has not encountered for some time and so forgotten' (73e.1–2).

a4) One can be reminded of Simmias by a drawing of Simmias (lit. 'Simmias drawn'). Thus one can be reminded by objects both like and unlike that of which one is reminded' (73e.9.ff). When the discussion moves on to the question of being reminded by similars, the examples are pictures. All the other examples which Socrates gives of being reminded seem to be examples of being reminded through associations.

a5) 'In any case of being reminded by 'likes', one must realize that what reminds one falls short of that of which it reminds one' (74a.5–7).

b1) 'If someone is reminded of the Equals themselves he must previously have know the Equals themselves.'

b2) Socrates demonstrates the applicability of a2 to the equals from 74a.9 to c.10. Socrates tries to show that our knowing equal things does not account for our knowing the Equals themselves or the Equal itself. Socrates states the general form of this proposition as a condition for being reminded at 73c.4–d.1. At 73d.3–9 he gives an example: to know a lyre is not to know a man, and so if a lover on seeing a lyre of his beloved thinks of his beloved he must already have knowledge of the boy before seeing the lyre. His seeing the lyre and his taking the form of his beloved in his mind's eye are contrasted. Since the form of his beloved is not the lyre nor a part of it, knowledge of the former may never be derived from knowledge of the latter. This is obvious, but in the case where one of the objects of knowledge is a likeness of the other, as are equal things of the Equal itself, it is not so obvious that a knowledge of the second cannot be derived from a knowledge of the first. The passage from 74a.10–c.9 is an attempt to prove this unobvious point in the case of the equals and it is a point which is vital to Socrates' argument. Socrates must show, not that equal things differ from the Equal itself, but that a knowledge of the Equal itself cannot be derived from knowledge of equal things.

How do we interpret 74b.7? 'Do not equal stones and sticks sometimes, though the same, appear equal to one person, but not to another?' Or "do not equal stones and sticks sometimes, though the same, appear equal to one thing but not to another?' Our second interpretation of 74b.7 is the one more likely to say what Socrates meant. For the first interpretation is not obviously to the point. The fact that some equal sticks appear unequal to some people is not an obvious barrier to a man's deriving his knowledge of the Equal itself from them. On the second interpretation 74b.7 refers to single objects, all of which are unequal to some things if equal to others. Socrates is arguing that from the mere perception of single objects nobody may derive a knowledge of the Equal itself, since all single objects in their relations to other objects are unequal as well as equal. We cannot learn of equality from any number of sets of equals because each member of each set is unequal to some things while equal to others.

Simmias admits this reasoning. So this argument demonstrates to his satisfaction that the case of the equals satisfies the condition specified in a.2. The satisfaction of this condition establishes any case as a case of being reminded. Hence the argument establishes the case of the equals as a case of being reminded. And so Socrates insists immediately after at 74c.13 – d.12. 'It must be a case of being reminded.'

b3) 'Being reminded by equal objects of the Equals themselves is an extreme case of being reminded, a case of being reminded of those things which one has not encountered for some time and so forgotten.' b3 is claimed by Socrates at 75e.2–7.

b4) 'To be reminded of the Equals themselves by equal objects is to be reminded by likes.' This claim is never made in quite this way, but is an element in Socrates' statements at 74e.3, 75a.2, 75b.1, etc.

b5) Socrates asks at 74d.5–7 'Do they [the equal sticks and stones] appear to us to be equal, in the same way as the Equal itself? Do they or do they not fall short in some respect of being such as is the Equal itself?' Precisely what has Socrates in mind when he asks this question and what Simmias when he replies emphatically 'They fall short by a long way'?

In establishing proposition b5 of my analysis Socrates and Simmias are not talking of any difference between equal objects and the

Equals themselves; they are talking about a single psychic event which occurs at a particular time under certain special circumstances. This psychic event is the realization of a difference or differences between equal objects and the Equals themselves. The problem is what difference or differences?

There is an oddity about both of our interpretations of 74b.7 if they specify this respect in which equal objects fall short of the Equal itself. Socrates and Simmias agree that when they see equal things they realize that the equal things try but fall short of the Equal itself (74d.7–10). So when Socrates and Simmias see equal things they think either 'These things are equal to some things and unequal to others' or 'These things are equal for some people and unequal for others'.

Now speaking for myself, I can say that neither of these thoughts comes to mind every time I see things as equal. Of course it may be that I just differ from Socrates and Simmias in this respect. It may also be that since the realization that equal things fall short of the Equal itself is so obviously not a part of our ordinary experience, it does not matter if our interpretation of that realization does not make it commonplace.

Again, 'falling short' has already been mentioned in the preliminary exposition. There it is a feature of the case of being reminded of Simmias by his picture. Simmias agrees that in such a case one must see that the picture falls short of its original. How? Perhaps in that it is just a picture, a two-dimensional, static representation on a wall. This difference is quite unlike the differences which Socrates and Simmias have discussed in the case of the equals. For in the case of the picture there can be no equivalent to the difference between the Equals themselves and the equal objects mentioned from 74b.7 to c.5. For that difference depends on the opposites, equal and unequal, to which there is no analogue in the case of the picture.

In the case of the picture, the difference which one must realize between what reminds one and what it reminds one of may be the difference between Simmias himself, just Simmias, and Simmias drawn, Simmias on the wall. And, of course, this way of expressing the difference employs Plato's famous idiomatic language of the x itself by itself, language used throughout this argument in the

Phaedo. Hence the respect in which equal sticks must be seen to fall short of the Equals themselves or the Equal itself may be merely this, that they are seen to be sticks, too, just as the picture of Simmias is also paint and wall.

Equal things fall short of the Equal itself because as well as being equal, they are 'mortal trash', in Socrates' phrase from the *Symposium*.[1] Anyone who sees them must see that they are not just equal, that is, not that something which is the Equal itself. I am not saying that every time I see things as equal I think to myself 'These things are like the Equal itself but fall short of it in such and such a way' any more than that every time I recognize a picture of a friend I rapidly enumerate all the differences between picture and friend. But there is a more or less unconscious realization there, which must be there if I am not to mistake the picture for the friend. This realization is of those differences which I have described in my account of its 'falling short'. But so rapid usually is the process of this realization that it cannot easily be put in the form of propositions.

In the *Republic* there is another explanation of how an image differs from its original. The picture of a bed pictures only one aspect of the bed out of the many aspects under which the bed appears to us. But that single aspect may be so well pictured that we take the picture of the bed for a real bed. In its turn the three-dimensional bed may have been for Plato just one version or aspect of the idea of the bed in the mind of God. This is how any instance of an idea falls short of its incomparably richer original. However perfect, it is partial. Likewise, any set of equal sticks or stones, however equal, shows only one aspect of the Equals themselves, and no instance can ever do more.

Socrates, we may feel, should show that those analytic and psychic features which are essential to a case of being reminded by a picture are to be found in the case of our perceiving equal things. But this he does not show. He argues only for the appropriateness of one of the essential features to the case of perceiving equal things—that the knowledge of the Equals is not to be derived from a knowledge of

1. *Symposium,* 211 c.3

equal things. At this stage of the argument at least, he has not shown even that when I perceive equal things I think of the Equals. Nor does he show it. He merely asks Simmias whether it is from the equal things that Simmias has taken his knowledge of the Equal (74c.7–9). Simmias immediately agrees that he has.

This lack of argument has some dramatic justification in that Socrates' audience requires not a proof but a reminder. And so the course of the exposition is not so much a reasoned proof of each stage of a complicated argument, as the proposing and accepting of the correspondences between the points of the general theory of recollection and the case of perceiving equal things. A3 and 4, for example, are never shown to be appropriate to the case of perceiving equal things, though they are often assumed by Socrates to be so. In the course of the exposition a1 and a5, the 'falling short' condition, are immediately accepted as appropriate by Simmias. Socrates only has to ask.

So this exposition of the theory of recollection is more like Socrates' account of the correspondences between the Sun and the Good in the *Republic*. It is not at all like a proof.

I will now ignore the structure of the *Phaedo* argument, and examine instead Socrates' understanding of how we come to see things as equal. I am more interested here in Socrates' theory than in his arguments for it. In fact, as I have suggested, this passage in the *Phaedo* is not much of an argument. It is more like a list of correspondences between 'being reminded' and 'seeing equal things'. These correspondences are accepted by Simmias without much discussion, and the theory thus developed is shown to yield the required conclusion—that the soul existed before birth. But the argument is not convincing to anyone who does not accept many other Socratic theories. The argumentation is *ad hominem*, in that it relies upon a wide and unusual agreement between questioner and respondent. Hence the argument will not be understood merely by examining its logic: to understand it we must try to see what in the process of seeing things as equal could possibly correspond to the process of being reminded.

Socrates proposes that when someone sees the equal sticks and stones he is reminded of the Equal itself and realizes that the equal sticks and stones fall short of the Equal itself. Therefore, he suggests, we must have known the Equal itself before that time when we first saw that equal objects fell short of the Equal itself (74e.9–75a.3). This may be paraphrased by 'Whenever someone sees some objects as equal, they are reminded of the Equal itself, and realize that those objects differ from the Equal itself in not having properties which the Equal itself does have or vice versa.' So far the only component of original statement and paraphrase which is readily acceptable is that people do see objects as equal.

Unlike Simmias I have no view about the Equal itself. So I shall work from the picture analogy. Suppose then that I agree that if someone is reminded of something by something else, then they must have known the thing of which they are reminded before being so reminded; that people see some things as pictures of Simmias, that a picture of Simmias is always seen to fall short when it serves as a reminder of Simmias himself; and that there is someone who is Simmias. Suppose that I accept Socrates' model of recollection in its most general statement from 73c.1 to 74a.7. Must I agree that anyone who recognizes a picture of Simmias must have known Simmias himself first?

The obvious counter-example is recognizing a picture of a film-star without ever having met the film-star. This can be done by people who have seen other pictures of the film-star and who see a likeness between this picture and those others. I am not interested here in how they come to know the name of the person pictured. In these cases they do not recognize the picture as being of someone whom they have met, but only that it is a picture of the same person of whom they have seen those other pictures. But they can be as sure of this as they are that some picture is the picture of their closest friend. And so we may reply to Socrates 'Certainly one way in which I come to be able to recognize a picture of Simmias is through meeting Simmias himself. But, as described above, there are other ways which do not require that I meet him. And so with equal sticks and stones. I have seen many equal things and from them have constructed a notion of equality and the forms that it may take, and I have never met the Equal itself, whatever that may be.' And so in one commentary on the *Phaedo*: 'Probably most people nowadays, if asked where we get such conceptions as perfect equality or perfect straightness from, would reply that they are derived by a process of 'abstraction' by comparison of a series of particular instances.'[2]

Here is another difficulty. Seeing a picture of Simmias as a picture of Simmias does not involve any visual perception over and above those had by someone who does not recognize the picture as being of Simmias at all. This is also true of seeing equal things as equal. Someone may or may not; and if at the time they do not, they may later recall them as equal. But if seeing objects as equal involves no visual perception other than those of just seeing them, what tells me that they are equal?

If I ask the same question of my seeing the picture as a picture of Simmias I answer 'The picture jogs my memory in some way' and then I can point to the eyes or the expression of the man in the picture and say 'They are just like Simmias'. But this only after I have recognized whose picture it is. I may use the phrase 'to construe the features of the man in the picture as the features of Simmias' to describe part of the process of being reminded. It is as though I read a memory into the picture. But not uncritically, for I can stop and ask myself 'Does the picture in fact bear the imposition?' Some pictures, a good forged passport photograph for example, can bear several different impositions—'It's got a look of all sorts of people'.

Recognizing the picture is seeing it in a certain way and so also with seeing objects as equal. In recognizing the picture I impose a memory upon it, but to this I can introspect no equivalent in seeing objects as equal. I compare the objects carefully to one another, but I cannot find in myself any further item which I use to examine whether or not they are equal. The theory of recollection is most strange to us for our not thinking of equality as an object known to all. To see two objects as equal I compare them and see that in some or all dimensions they are alike. This is quite unlike recognizing a picture of Simmias. Here an additional factor, my memory of Simmias is also at work. But Socrates often talks of the ideas as like the

2. R. Bluck, *Plato's Phaedo*, p. 63.

things we see and our knowledge of them knowledge by acquaintance. This is the sense of *episteme* throughout the argument. It is so used even of the lyre and the man in the preliminary examples (73d.2).

These two objections to the theory of recollection are answerable. And those who suppose that Plato means, more or less, the 'abstract concept of equality' when he talks of the Equals themselves or of the Equal itself, are, I think, mistaken.

How do I acquire the ability to see objects as equal? For Plato this ability is closely linked to being reminded of the Equal itself. This is a necessary and prior condition. To those who say that we derive our concept of equality by abstraction from the observing of a series of many equal things, Plato may reply 'That may be so but the Equality of which I am talking must come to mind before we can see anything as equal. We cannot learn of this equality from contemplating equal things for our knowing it is already part of our seeing them so.' Two questions: why can I not learn to see things as equal from looking at things? And why, if I cannot, is acquaintance with the Equal itself and its recollection the correct explanation of my ability?

The answer to both questions is given by the distinction between the Equal itself and the equal sticks and stones which Socrates makes from 74b.7 to c.5. Every set of equal objects which a teacher of equality may present to his pupil is also a set of unequal objects. The teacher cannot direct the attention of his pupil to their equality, for it is no more a feature of any single object than its opposite. How then does the pupil ever learn to see things as equal?

Seeing equal objects in the world is not by itself sufficient for seeing them as equal. We can imagine a man who has never seen things as equal in the world. Just seeing things which need not be seen as equal is not sufficient for our seeing them as equal. But perceiving things which cannot be perceived without being perceived as equal is sufficient for our perceiving them as equal. And this I think is the distinctive feature of the Equals themselves. If we perceive them at all we must perceive them as equal for that is all that is to be perceived. And once the knack has been acquired, we may repeat the process indefinitely even in those cases where we need not. The perception of the Equals themselves enables us to see other things as equal.

To return to the case of the man who recognizes the film-star's picture but who has never seen the film-star in person. I have argued that to recognize is to be reminded and to be reminded is already to know. What does the man who recognizes the pictures of the film-star already know? If he had no other knowledge of people or of himself, but had been confined just to pictures of one person, he would know neither that these were pictures of a person nor that they were pictures at all. To see for himself that they were pictures of a man he must know how a man looks. Otherwise the picture would seem no more than an item of strange graphic design, and not a picture of anything at all. But knowing how men look he can infer just from the picture of a particular man how that man looks, and when he recognizes other pictures of that man, the image which he imposes on those pictures is the constructed image of that man, not of a picture of the man. And so he sees that the pictures fall short.

The film-star is a member of a class all of whose members share or copy the characteristic 'being a man'. And that characteristic is such that we may know it without knowing all the members of the class. But the Equal itself is more like 'being a man' than like an individual man. The film-star is not just a film-star, but also a man. But the Equal itself is just the Equal itself. Another phrase which Plato uses is the Equal itself by itself. But if it is just equal, it can have no characteristic in common with anything other than itself through perceiving which and seeing the images of the Equal itself in the world I may construct an accurate idea of the Equal itself. For the Equal itself is terminal; by definition it can only belong to the class of equal objects and it cannot belong to any other class whose characteristic features it shares. But Simmias belongs both to the class of 'Simmias objects' (himself and his images) and to the class Man. And so we may abstract an idea of him just from his pictures.

There are other problems. Socrates distinguishes the Equal itself from equal sticks and stones by claiming that the Equal itself is never unequal. But is not the Equal itself unequal to everything other than itself? Or are not the Equals themselves unequal to everything other than another of the Equals themselves? Socrates assumes not. The language helps him here since it makes a sort of sense to claim that the Equal itself or the Equals themselves by themselves are just equal and nothing else whatsoever. But then we ask 'But are they not also singular or plural? Are they not also ideas? And what is this equivocation between the Equal itself and the Equals themselves?' There is a certain primitive force to the expression 'the Equals themselves'. These Equals, whatever they are, are busy being equal while our term 'equality' is just a name.

Socrates explains to his disciples that our knowledge of the Equal itself is recalled every time we see things as equal, and that this knowledge is not acquired in this life but is somehow constantly with us. The task of the philosopher is to fix the mind on and understand ideas such as the Equal itself. Plato is trying to make clear to our conscious minds those processes of our thinking of which we are but partially conscious.

These processes are continuous. They are operative throughout all our lives and thoughts. But they are not clearly understood by us because they operate somehow in an area to which we do not give our attention. They are not themselves visible but they effect our understanding of what we see—the understanding without which we cannot think or speak about what we see. To furnish any thought, to see any thing, we must use such categories as one and many, same and different, equal and unequal. The ideas are constantly operative in our thinking, even though what we see may so preoccupy us that we pay no attention to them. It is their constancy that qualifies them to be 'the (things) that are being', to translate the Platonic phrase usually translated 'the real things'. This substantive use of the present participle of the verb 'to be' conveys the presence and constancy of these ideas to our minds. So present and constant are they that we may fix our intellectual eye on them as long as we wish. In this they contrast with the rapidly changing world through which we move. And so we may have unchanging knowledge of them for what we discover to be true of them now will be true of them always. They are the unalterable elements of our thinking. The sights and sounds around us change but the ideas are forever the same.

We know and we do not know. We have crossed a stream which divides the kingdom of our minds in two and we have forgotten the other side. The ideas are there but we do not see them. Perhaps it is because they are so familiar to us that we do not notice them. Plato thought that everybody in the world already knew what was to be known: what they needed was the way to knowing that they knew.

This passage from the *Phaedo* is not a proof of immortality. As I have said, it hardly constitutes an argument at all. But it is perhaps a demonstration of immortality. It is the practical demonstration of an understanding of the processes of human thought, which convinces its exponent at least that there is that of him which does not change. This conviction allows him to await his appointed death in calm. By fixing his mind upon the ideas he prepares for the experience which he expects after his death and ensures that his transition will be as easy as may be.

3

The Divided Line

Republic 509d–511e

THE FOLLOWING two chapters are studies of the middle *Republic*, studies of the Divided Line, the Cave, and the Good. Socrates opens this part of the discussion with his account of the Sun and the Divided Line. Here I am with Glaucon: Sun and Good are too difficult. But even Glaucon manages to recapitulate most of what Socrates says about the Divided Line.

On the Line Socrates locates four levels of human understanding. But in the passage considered in this chapter, from the introduction of the Line to the end of Book VI, he explains only two of these levels, *noesis* and *dianoia*. These two levels are the most intelligent and so his account of them is the most intelligible. His account is scientific, is itself the account which proceeds from the understanding which Socrates called *noesis*. We do not have to go far for an illustration of the method. The Line too illustrates how a visible image may help us to see an intelligible proportion, a major point in the distinction which Socrates makes between *noesis* and *dianoia*.

Each section of the Divided Line represents one of four levels of human understanding. These four levels of thinking are bound together in these proportions. The first use to which Socrates puts the Line is to locate upon it sets of entities. Since there is controversy concerning the identification of these sets, I had best begin here. Two sections of the Divided Line are for 'the visible kind' (lit. 'what is being seen') two for the 'intelligible kind' ('what is being thought'). Of the two sections of the visible, one is for images of visible objects, such as shadows or reflections, the other for what these image (509e.1–501a.6). The problem is what are the corresponding entities for the sections of the intelligible? The problem arises from the way in which Socrates introduces these intelligible sets. For at 510b.4 his interest suddenly shifts from entities to modes of understanding; and just when the reader is prepared for the intelligible entities he is given a complex epistemology. Socrates remarks that geometers and mathematicians use visible diagrams in order to aid their thoughts about the Square and the Diameter and so on. These diagrams are images. So the entities of the two upper sections of the Line may be at the top ideas, the Square itself for example, and in second place diagrams of ideas, the drawing of a square for example. This provides us with the following equations: the Square itself is to the drawing of a square as is a visible object to its reflection and as the Square itself and the drawing of a square are to a visible object and its reflection.

The first of these equations, the Square itself is to a drawn square as is a visible object to its reflection, is of the same order as the one which underlies Socrates account of recollection in the *Phaedo:* the Equal itself is to equal things as Simmias to his pictures.[1] In the *Phaedo* Socrates talks of pictures, here of visible reflections. The difference makes little difference since reflections are also only recognizable in the light of their originals. But here in the *Republic* there is one important refinement, for the second and third terms, drawn squares and visible objects, may be the same, since a drawn square is also a visible object. And so in Socrates' exposition the same object, a drawn square, may occupy two sections of the line, for it may be viewed either as the image of an idea or as a visible object.

The last item of the complex equation is odd: its presence indicates that the visible mode of reflection between visible objects and their images is itself an image of the intelligible mode of reflection between ideas and particular objects. And so we see that the visible world is so perfect an image of the intelligible world that it even contains within itself an image of its relation to that intelligible world.

It will be objected that this interpretation is impossible since it assigns to the lower section of the intelligible visible diagrams, and

---

1. *Phaedo*, 73e.

visible diagrams could not be said to be intelligibles. To this I reply that the only images Socrates mentions in the context of his analysis of the intelligible are these diagrams, that he calls them images just as he calls shadows and reflections images (510b.4), and that Socrates himself no less than three times (510b4; e.3; 511a.6) remarks that these diagrams were the originals in the divisions of the visible kind but are now to be thought of as images, presumably when considered under the aspect of the intelligible. In the last chapter I have tried to show how physical objects that partake of ideas can be viewed under two distinct aspects, as 'equals' and as 'sticks'—that their partaking of ideas makes them to some degree intelligible. If however we do not accept that visible diagrams are a part of the intelligible then we are left with a gap and no indication from the text as to what should fill it. Hence the introduction by some scholars of a special range of mathematical entities, for whose existence there is, as far as I know, no other certain evidence in Plato's work.

From 510b.4 Socrates' exposition is no longer a progression of points but a recrossing of the same points several times. It is easiest therefore to consider all that is said under the two headings *noesis* and *dianoia*.

Noesis

In respect of the intelligible, geometers and mathematicians are contrasted with philosophers throughout the next paragraphs: geometers and mathematicians postulate the Odd and the Even and so on, give no further accounts of these postulates, and immediately proceed to the deductive proofs of their demonstranda; the philosophers on the other hand do not use these postulates as principles from which to deduce other propositions but use them as stepping-stones or spring-boards to grasp that which requires no assumption and is the starting point of all. So much Plato himself tells us (510c.1–d.3). The problem is to know what the philosopher does and what it is he grasps. My answer to the problem is this: the starting-point of all, which requires no further assumption, is just such a being as Parmenides describes in the 'Way of Truth'. I propose that Plato in describing the grasping of the unhypothesized beginning is describing essentially the same experience as Parmenides describes in the 'Way if Truth'. There are similarities between the work of the two philosophers: the language of the Line is in some important elements identical to the language of the 'Way of Truth', especially in the value given to *einai,* to be, and the verb so often correlated with it, *noein,* to think; second, both Parmenides and Plato teach a philosophy of enlightenment, of an extraordinary experience which makes certain sense of the world; third, both describe the experience as the highest fulfilment of desire; fourth, the structures of the Platonic and Parmenidean systems are similar in having some ultimate premiss, the 'unhypothesized beginning' or 'is' which requires no further support, but guarantees the remainder of the system. These similarities are so far as I know confined to Parmenides and Plato. No other philosophy of the time shares them. But, of course, there are a number of important differences between Plato and Parmenides also. Nonetheless, for Plato as for Parmenides being is thinking. The philosopher, in grasping this principle, understands that his being is his thinking and that the ideas postulated by mathematicians are indeed intelligibles, *noeta,* the fixed terms of his thinking. He sees that of this thinking the ideas are parts, and that their differences are reconciled in the thinking which embraces them all. This understanding requires no further justification because it is already as certain as can be, and it is this because it is the understanding of what is clearest to our minds—our own thinking.

The starting-point as it appears on the Socratic Line is such a being as is described in Parmenides' poem. *Noesis* is the ascent to and descent from the experience of that being. The terminal points of ascent and descent are being and ideas. In the descent the ideas, which depend from being, are now fully understood in the light of that being. In the ascent they are used as spring-boards to the final vision. The word here translated 'spring-boards' suggests not a slow process of analysis but a sudden intuition, a leap not a climb. This is one of Plato's few metaphors in the Line. Nonetheless it is by the power of dialectic that such a leap is made, and it is the *logos* itself which grasps being (511b.4). Likewise it is to the judgement of the *logos* that the goddess presents her poem the 'Way of Truth'.[2] In the

Sophist Plato writes and the Elean stranger, from Parmenides' city, says:

> And the man who can do dialectic discerns clearly one form everywhere extended throughout many, where each one lies apart, and many forms different from one another but embraced from without by one form, and again one form connected in a unity through many wholes, and many forms entirely marked off apart.[3]

It is the realization of the necessity of such a system that it must be so, that is in part the grasping of the beginning which requires no further assumption.

Ideas are intelligible only in so far as they are contained by intelligible being, from which it follows that no idea can be known as distinct from being or from any other idea. They are intelligible only within the system, and must be connected to each other by the same necessary logical relations as, say, the notes of a key-board. Socrates says in the *Meno*:

> All nature is akin and the soul has learned everything, so a man who has recollected a single thing is in no way prevented from discovering everything.[4]

Socrates does not argue his claim; for he knows that what is to be known must be so.

In the *Philebus* Socrates gives a description of the establishment of the science of letters by the god or godlike man, Theuth. Theuth observed that the sound of speech had had no bounds set upon it, and proceeded to make distinct classifications according to differences between the elements which composed that sound. These elements are the letters, and Theuth saw that no one could learn any one letter of an alphabet by itself without learning the others. Thinking that this was a common bond which made them in a way all one, he assigned to them all a science and called it 'grammar'.[5] The similarity between this description and the passages just quoted

2. *Parmenides*, Frag. 7.
3. *Sophist*, 253d.5-e.2.
4. *Meno*, 81c.9-d.4.
5. *Philebus*, 18b.6-d.2.

from the *Meno* and the *Sophist* indicates that this mode of conceiving of a unity within a plurality is a common thread running through the dialogues. In the *Epinomis*, after an account of mathematical studies, there is this passage:

> Him who has mastered all these lessons I account in truth as wisest: of him I dare affirm—it is a fancy yet I am in earnest with it too—when death has ended his allotted term, if he may be said still to endure beyond death, he will no longer be subject, as he is now, to a multitude of perceptions; he will have only one allotted portion even as he has reduced the many within himself to one, and in it will be happy, wise and blessed all in one.[6]

For Socrates the experience of being as thinking, *noesis*, is the beginning of all certain knowledge. For him as for Parmenides this experience is the highest inspiration of which we are capable. Nothing else can be understood but in relation to this experience and the Socratic theory of this relation is most elaborate. The experience ensures that thinking is being, and is unitary, unchanging and so on. The problem is to relate this to our varied experience of the world without impugning its unity, homogeneity, immutability; but to relate it in such a way that our other experiences also remain in some sense valid. In the Line are the foundations, the first stages of Socrates' method of relation. He makes the experience of being the perception of a single aspect of a complex system of eternal unchanging ideas. *Noesis* is the grasp of the system as a whole, as one. *Noesis* teaches the philosopher that the ideas to which his studies have accustomed him are not discrete nor independent, but inseparable parts of a complex whole, embracing the entire range of the intelligible and all his studies at once. He understands that of this whole he has always been dimly aware, and that this awareness has allowed him to comprehend by intellect and diagrams the necessary relations between the entities of his studies.

Now through the experience of this being he comes to see the point of that education. And that point is the science of all the ideas and all their relations, the complete and perfect science which

---

6. *Epinomis*, 992b.2–8.

unites all the branches of mathematics, music and astronomy, and understands all of them completely. The philosopher who has passed through all these subjects in his training now reaches a point from which he views them all whole and at once. He now looks down upon them from a point beyond them and above them; and only now is he able to map them properly, knowing that they comprise a single landscape. His previous maps of parts of that landscape, made during his ascent to the summit, can now be incorporated in a greater whole. He understands the aim of his laborious education. But he learns not by some further study of the same sort as those previous studies, but through an intuition.

By itself this intuition does not show the philosopher exactly how the ideas are arranged in relation to each other. Rather it shows him the necessity for that arrangement. The philosopher sees this in seeing that he and his world are no more than the sum of his experience, his thinking, and that thinking must be continuous, complete, and such that each part of it is somehow present to every other part. How this last may be a feature of the ideas we have already seen. The grasping of the unhypothesized beginning is the realization that being is thinking, and the understanding through this realization that all the ideas are equally necessary to the whole and completely interrelated.

Dianoia

Dianoia, understanding, is the name which Socrates gives to the intellectual operation of some geometers and mathematicians of his day. *Dianoia* is between *noesis*, or true science, and opinion. It needs no more description than Socrates gives it; we recognize the study of geometry as practised in the school-room. Each entity of the system is taken for granted, its representation on the blackboard easily grasped, and its demonstrable relations to the other entities of the system the only object of study.

Geometry and mathematics, as so practised, will never be true sciences, since their postulates will never be as clear and therefore as sure as is being. Nonetheless the Odd and the Even, the Square, the Angle and so on are its forms, so near to this being that even those who have not grasped its nature may appreciate the necessity of the connections between them. Unlike the philosopher and dialectician, mathematicians and geometers use visible symbols to aid them in their task.

It may be objected to this interpretation that it allows of no distinction between *noesis* and *dianoia*, since both now deal with ideas. But *noesis* and *dianoia* are not so much distinguished by being correlated to different sets of entities as by their different approaches to the same set. *Noesis* examines ideas and only ideas, while *dianoia* studies ideas through the medium of visible diagrams; *noesis* understands ideas for what they are while *dianoia* proceeds merely to conclusions which relate ideas in certain ways. Of course, on this interpretation the strict one-to-one correspondence between the four sets of entities and the four *pathemata* of the soul (511d.6–e.4) breaks down, since *dianoia* is now correlated with two of the sets of entities, ideas and visible diagrams. This does not seem to me a very serious divergence since there are still four sets of entities, four *pathemata* in the soul, and it is still possible to distinguish *noesis* from *dianoia* by reference to different sets of entities, *noesis* being concerned with ideas only while *dianoia* deals with visibles as well.

Socrates' remark at 511a.1 that mathematicians seek to see those things themselves which cannot be seen but by *dianoia* by using images indicates that there are indeed two ways of thinking about visible originals. First they can be thought of simply as visibles, just as the equal sticks can be seen merely as sticks, and in this case they occupy the third place on the original Line. Second, they can be used as images 'to see those things themselves which cannot be seen but by *dianoia*,' namely ideas, in which case the visibles are used as images in their intelligible aspect and occupy second place on the original Line. The care with which Plato makes this point suggests that he was fully aware of this ambivalence in the status of the visibles and was trying in his account of *dianoia* to be as clear as possible.

Plato's account of *dianoia* has two important elements in common with Kant's account of mathematics. For Kant, too, visualization was essential to the work of geometers, an intuition of space. In arithmetic, according to Kant, an intuition of time was essential. For Plato, the dianoetic mathematician uses diagrams, pebbles to stand for numbers, lines in the sand with the slave boy in the *Meno*, and they enable a sharp distinction between this kind of mathematics and what the noetic mathematician does: the noetic mathematician makes no use of such diagrams but proceeds from ideas to ideas and ends with ideas. For Plato as for Kant the use of diagrams, the necessity of *anschauung,* was definitive of mathematics as practised.

Again, for Plato as for Kant, the pure intuitions of such mathematics and the space and time in which they occur must precede any world experienced through the physical senses. Since mathematical intuitions of space and time are prior to and shape physical space and time in our lived experience through the senses, the physical realm of nature is a projection of the mathematical realm. This is exactly how Plato represents it in the relations between the objects of *pistis, eikasia* and *dianoia.* For Plato as for Kant that is why mathematical models must always apply to the physical world. And this, of course, is the explanation for the unreasonable effectiveness of mathematics in the natural sciences.

The space and time with which we organize our experiences in the natural world are understood more purely by the inward mathematical powers of the mind. The space and time of nature are mathematical space and time organized into a single order in which every item has a spatial or temporal place in relation to every other item. This is not true of mathematical space and time, according to Kant. The mathematical forms, equations and demonstrations apply to physical space and time not by serendipity but because physical space and time are themselves organized entirely by the mathematical ideas in the first instance. In all this Plato and Kant would agree. And they would agree as philosophers of mathematics, not as mathematicians. For it is outside the boundaries of mathematics as practised to propose and argue these truths. To this extent Kant's analysis is noetic.

But for Plato, unlike Kant, the mathematical ideas may be comprehended in an entirely different way also. They may be seen in the light of the Good directly. The Kantian may well object here that there is no sense at all to be given to mathematical ideas independently of the pure intuitions of space and time. But this is clearly not Plato's vision when he carefully describes how the noetic mathematician makes no use of diagrams but deals only with ideas. It is almost as though Plato had read Kant and was explicitly contradicting him here.

Despite himself, Kant helps us to see just how great the difference is between *dianoia* and *noesis*, how divergent their procedures and methods. And in a beautiful squib Kant describes how Plato's belief in mathematical ideas beyond spatial or temporal intuitions is like a bird's thinking how much better it could fly if only there were no air to impede it.[7] To which Plato might reply that the silly bird is rather Kant than himself. For both philosophers, what ultimately supports them is necessarily without support. But Kant believed that this shows how reason must ultimately fail. Plato believed that at this point the reason is overwhelmed by insight, the only certainty, the clearest experience of all. From this point of view the mathematical ideas as a set have their own integrity, quite beyond the relations which *dianoia* studies. The apprehension of this integrity is the vision of the Good, of which Kant seems to have no notion.

The difference between *noesis* and *dianoia* is the difference between the noetic and the dianoetic mathematicians. It is the difference between the founder and the mere practitioner of a mathematical science. The exemplar of a founder here is Pythagoras, but Socrates too was believed to be the founder of the science of ethics. And both Pythagoras and Socrates were agents of Delphic Apollo. The founder of an art or science was an intimate of the patron divinity, as Triptolemus of Demeter or the Idaean Dactyls of Hephaestus. Neither in Socrates' account of *noesis* nor in mine is there much sense of rapture, though it is a governing theme in the account of the ascent from the Cave which follows. The only hint here is the 'leaps' or 'spring-boards' which the noetic mathematician makes of mathematical ideas such as even and odd.

The founder of a science establishes its terms and categories. Those who follow explore the interrelations between these terms and their applications. But it was a given that the founder had antic-

7. I. Kant, *Introduction to Critique of Pure Reason*.

ipated *in principio* all that the science would become in the centuries thereafter. To a degree it was possible for the mere practitioner to follow the founder even into the rapture. Socrates is clearly describing the process of an initiation in the Cave. He was entitled to make such revelations, it is said, because he had never been initiated in any of the mysteries and could break no oath of silence.

The founder of a science cannot, as a follower does, take the terms of the science for granted and proceed immediately to demonstrations. For the terms of the science are not then established. But founder and follower are practising the same science. The difference between them is that the founder intuits directly the entire science in principle. On the one hand, the founder's vision coordinates the terms and principles with the Good; on the other it looks on down to their unfolding through the future. The notion that the entire science is somehow implicit in the vision of the founder is a peculiarity of traditional thinking. And this same vision is actually shared by the initiate, who becomes from that time intellectually self-sufficient.

4

The Simile of the Cave

Republic 514a–519b

THE SIMILE of the cave begins:

> After this, I said, liken our nature to such a state as this in respect of education and the lack of it (514a.1).

The first four words are colorless. They throw all the emphasis of the sentence on to the word immediately after them—'liken', *apeikason*. This word is the key-word in the account of the Cave. In his account of the Line, as I have shown, Plato talks like a scientist; it is clear and undiluted. That account is introduced by the word *noeson* or 'contemplate' (509a.1). Though the Line has located upon it four *pathemata* he analyses there only the top two, and at the end of the Line he gives just the names of the third and fourth, '*pistis*' and '*eikasia*'. Thus the word *apeikason* at the beginning of the Cave simile echoes *eikasia* at the end of the account of the Line. In order to talk of these lower *pathemata*, the modes of thought which are most correlated with images, he uses an image, just as in talking of mathematics and science he has used the language of mathematics and science. Again, just as on the Line there are places for all the modes of cognition, so are there in the simile of the Cave. We may view our progress through Line and then Cave as the two parts of the same journey, and the alteration in perspective in our view of the same landscape is such as we would expect, had we seen all from the summit and now turned to look back at it from the depths. This change of perspective is exactly conveyed by Plato's change of style—he is talking of and we are seeing the same things in a different way.

Now the order in which Plato describes to us the features of the Cave. He begins:

Behold human beings as it were in an underground cave. (514a.2)

Then follows a general description of the Cave, the people, the fire, the wall, and the relation of the Cave to the outside world. That 'behold' is very powerful. It suggests that moment when a guide reveals to his companions an entirely new view. It is as though we stood with Plato on the very brink of the Cave, as though we had just stepped in from outside, and the whole of the inside of the Cave lay revealed to us for the first time. It is an overview. This initial vision lasts from 541a.2 to b.6. From 514b.8 to 515a.3 Plato describes to us the puppets carried along the wall. It is as though we had gone down into the Cave and he was describing to us what we passed as we proceeded further from the light of day to the Cave's deepest places. Finally at 515a.7 he gives us a description of the wall furthest from the entrance. We have at last reached bottom—our journey which began with the cutting of the Line is complete.

In this way the simile of the Cave is the tally of the Line. It is a new view of the Line, seen through an image. Once the primary division of the Line is crossed, we enter a world entirely comprised of images, so lacking in clarity that an image is the best or perhaps only way of describing it. The relations between artificial Cave and Line are intended to convey the relation Plato believes to hold between the worlds of opinion and intelligence. The Cave complements and images the Line—the two are so interwoven as to make a single work of art. Two ideas, one mathematical and the other metaphorical, seemingly quite disparate, are so combined that it becomes impossible to disentangle the second from the first and view it adequately just by itself.

There are two ways of coming to understand what Socrates means by the terms *pistis* and *eikasia*. The first is to deduce from the conclusions already given concerning *noesis* and *dianoia* what must be the analogous or proportionate natures of this other pair of the four modes of cognition related by the Line; the second is the interpretation of the simile of the Cave. In this chapter these two ways will be travelled simultaneously.

It is clear that the experience of the prisoner who leaves the Cave for the outside world represents *dianoia* and *noesis*. In the outside world there are three sets of entities which the prisoner comes to know. These are in ascending order the reflections or shadows of natural objects, natural objects, and the sun. The analysis of *noesis* and *dianoia* given above makes it clear that these three sets represent three sets of intelligible entities, which in ascending order are diagrams or symbols of ideas, ideas, being. In the Cave there are also three sets of entities which correspond to and represent the entities of the outside world. The three sets in the Cave are in ascending order the shadows of the models of natural objects, the models of natural objects, and the fire. If the simile of the Cave is precise, and if *pistis* and *eikasia* are analogous or proportionate to *noesis* and *dianoia* according to the divisions of the Line, then these three sets of entities in the Cave represent the correlates of *pistis* and *eikasia.*

From 517a.7–b.4 Socrates goes some way towards relating the entities within the Cave to the correlates of *pistis* and *eikasia.* He tells us that the Cave represents the world manifested through sight, and that the light of the fire represents the power of the sun. From these indications there is a temptation to conclude that the entities of the Cave stand for three sets of purely visible entities, namely images of natural objects, natural objects, and the sun. But this is to misunderstand the simile.

The strongest argument against this interpretation is that it renders *eikasia* most obscure. If *eikasia* is analogous to *dianoia,* then it consists in the understanding of one set of entities through the study of another set of entities which image that first set. The geometer learns about the Square itself with the help of diagrams of squares. So if the shadows in the Cave represent the reflected images of the visible world, then *eikasia* consists in the understanding of natural objects through the study of their reflections. But this method of study has never been anyone's unique practice either in Plato's day or our own. But at 515a.5 Socrates insists that the prisoners who see only the shadows represent the human condition in general.

The prisoners see nothing of themselves, of each other, of the object carried across the Cave but the shadows, and they think these shadows reality. In so doing they represent the human condition. I know of no one who thinks reflections and only reflections reality, but I know many who suppose the real world to consist of the physical objects around them. They call themselves materialists.

Let us hypothesize that the prisoners in bondage in the Cave represent materialists and that the shadows on the wall of the Cave represent what they call material objects. And let us suppose that Socrates is representing *eikasia* when he says:

> Now suppose that these prisoners had among themselves a system of honours and commendations, that prizes were granted to the man who had the keenest eyes for the passing shadows and the best memory for which usually came first and which second and which came together(516 c.7).

Eikasia then consists in the ability which some materialists have to predict the sequences and coincidences of material objects. It is like dianoetic geometry or mathematics insofar as it asserts and in some sense establishes connections between a range of entities without knowing the reason for these connections. But it is unlike geometry in that it is memory and not deduction which establishes the connection. In this respect the distinctions which Socrates makes between the doctor and the pastry-cook are interesting.[1] For Socrates attributes the ability of the cook to gratify palates to memory.

Two problems remain, to which there is a single solution. The first problem is this: if the shadows in the Cave represent material objects and the light of the fire the power of the sun, what do the models of natural objects, the third set of entities in the Cave, represent? The second problem is this: geometers use their diagrams to discover the properties of the Square itself and other forms; *eikasia* is analogous to *dianoia*; material bodies are analogous to the geometric diagrams: of what then do the materialists discover the properties by studying the sequences of material objects?

The material objects outside the Cave represent the ideas with which *noesis* and *dianoia* deal. In the Cave are models of those objects in the outside world. Clearly, then, *pistis* and *eikasia* deal

---

1. *Gorgias,* 464.

with entities which image the intelligible ideas of *noesis* and the mathematical sciences which constitute *dianoia*. The obvious candidates for the objects of *pistis* and *eikasia* are those kinds of the natural world which Socrates sometimes mentions, the Bee in the *Meno*,[2] or Man in the *Parmenides*.[3] These kinds produce material objects which belong to those classes which constitute the structure of the natural universe. And the materialist in his study of material objects classifies them according to these classes and produces rules or guides to memory which relate these classes to each other, and not the individual objects which he sees. But *pistis*, if it is like *noesis*, studies only these kinds.

The three sets of entities within the Cave are therefore capable of at least two quite different interpretations: they may be taken to represent either shadows of material objects, material objects and the sun, or material objects, kinds of material objects, and the sun. Socrates himself says that the fire in the Cave stands for the power of the sun. But the sun has a dual power:

> You will say, I think, that the sun provides not only the power of being seen to the things that are seen, but also their becoming, increase, and nourishment (59b.2–4).

The sun not only gives light; it brings about the generation and development of what we see. 'The world manifested through sight' I take to be the entire world of becoming which we perceive with our eyes but must understand in terms of those processes which bring it to be.

To return to *eikasia*. *Eikasia* is the constructing of theories about the relations between entities of the visible world; these theories are based upon observation, upon the observed order of events. The visible entities so related are not considered individually but are classified in sets and the theories relate these sets. So general rules are constructed which allow the clever prisoner to guess what will happen on the basis of his past experience: an entity of type B will now follow this entity of type A, since in the past entities of type A

2. *Meno,* 72b.2.
3. *Parmenides,* 130c.1.

have always been followed by entities of type B. But the clever prisoner cannot be sure that what he predicts will happen, nor can he
say why it does even if it does. And so he can be said to rely on
guess-work, *eikasia*, rather than science.

The prisoner, the man engaged in *eikasia*, notes certain similarities between the shadows, the visible objects before him, and classes
them in different classes according to their similarities, giving
members of the same class the same name. But he does not understand why there are these similarities because he does not know of
the existence of the puppets, each one of which casts the many yet
similar shadows which he calls by the same name. He does not
know of the puppets because he cannot see them; they are not in his
direct field of vision. In this he is like the mathematician, looking at
visibles and thinking in terms of intelligibles. But he is even less
aware than the mathematician of the real nature of the terms of his
thought.

And it is here that *eikasia* differs from *pistis* just as *dianoia* differs
from *noesis*. For the man engaged in *pistis* is much more interested
in understanding the reasons for the common classifications than
in relating the objects so classified to each other. *Pistis* is looking at
the kinds and understanding how the visible entities which those
things 'cast' are similar by virtue of being cast by the same kinds.
This understanding is achieved not by looking at the far wall, at
what is to be seen or sensed, but by a fundamental turnabout, by
introspection into the nature of how we think about the phenomenal world. So *pistis*, even though it is part of opinion, is not primarily concerned with sights and sounds but with the origin of the
similarities between these sights and sounds. That *pistis* is not primarily of the phenomenal world is illustrated by the way in which
the puppets which cast the shadows lie outside, directly behind, the
prisoner's range of vision; that it is part of opinion, part of the
understanding of the Cave, is explained by the fact that it looks at
the origins of those similarities by which we understand the classes
of the phenomenal world. According to the image of the Cave, the
prisoner need only be freed and turn around in order to see the
puppets and understand their role. But at 515d.4–5 Socrates temporarily exceeds his image of the Cave to make it quite clear that the

way to understand the puppets is not merely to see them but to be forced to answer questions about them. Socrates' momentary departure from his frame of reference suggests that the puppets do not stand for visibles at all but for entities to which dialectic offers the only approach.

What are these kinds for which the puppets stand? They are not just classes even though they make possible the classifying of the shadows, for the relation between a class and one of its members is not the relation between a puppet and its shadow. Somehow the visible entities which belong to a certain class are, for Plato, no more than images of the kinds for which the puppets stand. And just as the puppets are invisible to the bound prisoner so are these kinds to us though they help bring about what we can see. So in coming to understand these kinds we are freed from the error of supposing that only what we see is real: instead we understand that these kinds are far more real and constant even though invisible. The kinds like the mathematical ideas are elements of our thinking; they systematically fashion what we sense and are the forms in which we think about it. The unliberated man is only partially aware of these kinds and makes the mistake of supposing that only what is before his eyes is real. But *pistis* is the awareness that what we sense is but a projection of what we think.

This account of *pistis* is all right as far as it goes. But left here, it would involve me in an absurdity. For so far in my account of the four kinds of understanding I have given no place to Plato's own understanding of the world of becoming. For the Cave shows that Plato thinks of that whole world as an image of the intelligible, while neither *pistis* nor *eikasia* on my account realizes this. So there is the question: is either *pistis* or *eikasia* the understanding also of how the world of opinion images the intelligible world? It is clear that *eikasia* is not. Further, if *pistis* is the proper and full under-standing of the world of opinion as *noesis* is of the intelligible, then it must be an understanding of how the former images the latter. I judge that *pistis* is Plato's understanding and that it is not a stage in the prisoner's ascent as the simile of the Cave may perhaps suggest. *Pistis* is that vision of the world of becoming given to him who returns to the Cave from contemplating the ideas and being. The prisoner cannot appreciate that the puppets represent natural objects outside the Cave until he has seen these objects. For this reason if for no other the simile of the Cave follows the account of the Line.

Pistis is that overview, that appreciation of likenesses, with which Socrates introduces his simile. That first step into the Cave from the outside world represents best the understanding which is *pistis*. And I have argued that Socrates' account of knowledge and opinion are parallel and complementary in this further special way: in his account of knowledge he demonstrates *noesis* and describes *dianoia;* in his account of opinion he demonstrates *pistis* and describes *eikasia.*

5

The Model of the Cosmos

Timaeus 30c–d

THERE IS more that can be said of the statues or puppets since Plato talks of them in the *Timaeus*, in Timaeus' account of the creation of the universe. At 30c.3 Timaeus asks:

> In the likeness of which of the living things did the framer frame it (the universe)?

The answer he then gives is:

> To that whereof the other living things individually and according to their kinds are parts (30c.5–7).

What are the living things? Timaeus continues:

> For that contains within itself all the intelligible living things just as this cosmos comprises us and all other creatures.

For the identity of the intelligible living things, I can do no better than quote Cornford who writes:

> We have seen that, although the creator god, as such, is a mythical figure, the relation of likeness to model none the less subsists between the visible world and the intelligible. The model is not a piece of mythical machinery. The visible world, being 'in very truth' a living creature with soul and body, has for its original a complex Form, or system of Forms, called 'the intelligible Living Creature'. This is a generic Form containing within itself the Forms of all the subordinate species, members of which inhabit the visible world. The four main families, 'contained in the Living Creature that truly is', are enumerated at 39e: the heavenly gods (stars, planets, and Earth), the birds of the air, the fishes of the sea,

and the animals which move on the dry land. These main types as well as the indivisible species of living creatures and their specific differences, are all, in Platonic terms, 'parts' into which the generic Form of Living Creatures can be divided by the dialectical procedure of Division. The generic Form must be conceived, not as a bare abstraction obtained by leaving out all the specific differences determining the subordinate species, but as a whole, richer in content than any of the parts it contains and embraces. It is an eternal and unchanging object of thought, not itself a living creature, any more than the Form of Man is a man. It is not a soul, nor has it a body or any existence in space or time. Its eternal being is in the realm of Forms. Plato does not say, here or elsewhere, that this generic Form of Living Creature contains anything more than all the subordinate and generic and specific Forms and differences that would appear in the complete definitions of all the species of living creatures existing in our world, including the created gods. We have no warrant for identifying it with the entire system of Forms, or with the Form of the Good in the *Republic*, or for supposing that it includes the moral Forms of dialectic or the mathematical Forms, or even the Forms of the four primary bodies, whose existence is specially affirmed at 51b ff. Plato looks upon the whole visible universe as an animate being whose parts are also animate beings. The intelligible Living Creature corresponds to it, whole to whole, and part to part. It is the system of Forms that are, together with the Forms of the four primary bodies, relevant to a physical discourse, because they are the patterns of which the things we see and touch are sensible images, coming to be and passing away in time and space. [1]

Cornford adds in a footnote that the phrase which I have translated 'individually and according to their kinds' refers in the first place ('individually') to the 'forms of indivisible species' and in the second to the 'four main families which are enumerated at 39e'—Gods, birds, fish and land-animals. Hence Plato makes here a start on the task of producing a complete and hierarchical classification of the natural kinds, a task which he has not begun in his account of the puppets in the Cave.

1. F. W. Cornford, *Plato's Cosmology*, RKP, pp. 40–41.

I agree with everything that Cornford says in this passage. I would point out what I take to be an omission only, that we do have some warrant for supposing that Plato is referring to this same model of natural kinds in his account of the Cave and of *pistis*. Furthermore, this complete model which contains within itself all the intelligible living things may be paralleled with the sum of the ideas, according to the proportions of the Line. There are two distinguishable sets, one of mathematical ideas, one of intelligible living things, each complete and holding first place on one or other of the two major divisions of the Line.

The *Timaeus* throws more light on another obscurity in the account of the Cave. I have implied that Plato though that the power of the sun was responsible for the generation of living creatures on the earth and that the natural kinds mediated between the power of the sun and the sensible world so as to transform that power into creatures or creations belonging to the many different classes which comprise the structure of nature. At 41a.7 of the *Timaeus* God addresses the lesser Gods which he has made and tells them to make the other living creatures necessary to a perfect universe. These lesser Gods are the heavenly bodies, the earth, the stars and the sun. Cornford writes on this passage:

> This delegation of the rest of the work to the celestial gods may perhaps be connected with the notion that the heavenly bodies, especially the Sun, are active in generating life on the Earth. . . . In *Republic VI* the Sun is singled out among the heavenly gods as 'the offspring of the Good which most resembles his parent' He is the cause of the birth, growth, and nourishment of things in the visible world.(509b)[2]

It is noticeable however that in the *Timaeus* passage the sun is not so singled out.

So the *Timaeus* clarifies the obscurity of the Cave. The fire and the puppets reappear as the intelligible living things and the lesser Gods, and in the same relations to each other. On this interpretation the simile of the Cave tells us no more than Plato himself tells

2. Cornford, op.cit., p. 141.

us more clearly elsewhere. But there is one serious difference: Timaeus' intelligible living things are called intelligible while the entities for which the puppets stand do not belong to the 'intelligible place' (517b.5). Socrates' position on the form of Man in the *Parmenides* lies between these differing positions of the *Republic* and *Timaeus*. He agrees that there are such forms as Man but not with confidence. Hence the status of these entities is unclear and so the differences on this point between *Republic* and *Timaeus* need not count for much against their similarity. The unclear status of these entities explains their relegation to the sphere of opinion in the *Republic*. Less clear than the mathematical ideas, the proper correlates of knowledge, they must belong to opinion if anywhere, to the likely story.

Such are *noesis, dianoia, pistis,* and *eikasia*—four types of understanding. Plato does not make it clear whether these are the only four. But he does make clear how the last three of these four derive from and depend upon the first, just as the correlates image or image images of ideas. All four attempt to bring system and unity to human experience. But *dianoia* and *eikasia* do not realize the presupposition, justification, motive and end of their attempt, the unity and order of thinking. Just as Simmias' picture is only recognizable to someone who already knows Simmias, so *dianoia* and *eikasia* involve some dim awareness of being, of its unity at which they aim. *Noesis* and *pistis* on the other hand are an exact recollection of that original acquaintance.

To see that two sticks are equal is to remember equality: the part brings to mind the whole. To remember equality is to conceive dimly of all the ideas of being: the part brings to mind the whole. In this way every thought of ideas, or of images of ideas, or images of images of ideas, leads back to and derives from the vision of being more or less clearly. Plato shows how every instant of our thinking implies our awareness of this whole and that our thinking in this sense does not change but is forever at one.

Imagine two super-powers at or near war. Let us call them Redland and Blueland. Redland is making a weapon of the greatest destructive power. The Government of Blueland hears of this and wishes to make an equivalent weapon if only to maintain the balance of power. To make such a weapon the scientists of Blueland must first know what it does.

Let us suppose that the effect of the weapon is to be an explosion the equivalent of the explosion of X megatons of TNT. Let us suppose further that there is in principle only one way of producing this effect. For the production of this effect requires the best possible use of the only material capable of producing it. For the weapon is as efficient as it can be.

If then Redland had indeed discovered how to make this weapon, they must have discovered the unique means of producing that effect. And if the scientists of Blueland by themselves and with no more information, also discover how to make such a weapon, they too must discover the unique means of producing that effect.

In principle there are two ways open to Blueland of discovering how to produce the weapon. The first is to steal the plans of Redland's weapon; the second is to develop an equivalent weapon for themselves. Let us suppose that Blueland's spies succeed in stealing what they think are the plans of Redland's weapon and send them to Blueland before her scientists have developed the weapon for themselves. In this case the scientists of Blueland would not be sure of these plans, of the rightness of the method in the plans. If these were indeed the plans, they would if they used them have right opinion, but they could be persuaded to drop this method and try another.

On the other hand, let us suppose that the scientists of Blueland attempt to develop the weapon by themselves. First, they must discover what the weapon is to do. Let us suppose that they somehow find this out. In that case, and with no more information at their disposal, they could theoretically come to understand how to make such a weapon simply by learning as much of the physical principles utilized by the weapon as have the scientists of Redland. If they acquired more information about Redland's weapon that would be a bonus but they could do without it. And if they succeeded in discovering the principles, the only way of employing them, and that this was the only way, they would then be in a better position to assess reports about Redland's weapon.

The *Timaeus* shows exactly what that effect is which the Craftsman produces. It therefore provides the original of one of the images in my war-story—the original of the weapon's effect. Timaeus is quite clear that the Craftsman frames the physical world as he does in order to produce one effect above all—that the physical world be as like being as it can be. Timaeus goes through several of the attributes of being as Parmenides described it and shows how the Craftsman made the world accordingly. It contains all the kinds and all the elements in a perfectly integrated unity. It is intelligent, complete, solitary, self-sufficient, globular, unageing and perfect. In the case of the weapon it was necessary for the scientists of Blueland to know at least what effect they were to produce. And so also the Platonic scientist becomes aware of what it is that the Craftsman makes of the world through a knowledge of the being which Parmenides describes, in the absolute unity of the mathematical ideas.

Passages in the *Timaeus* correspond to my war-story at other points. The Craftsman is limited in his material which must be corporeal and visible and tangible.[3] Likewise in our story we may imagine that the scientists of Blueland know that only certain materials are available to Redland. Again, Redland's weapon makes optimal use of the material available, and the best use of those physical principles necessary to the production of the effect. Likewise Platonic scientists work on the *a priori* assumption that the Craftsman makes the best use both of his material and the possibilities available to him. And again Platonic scientists hope that their knowledge of the material, the possibilities, and the intended effect will enable them to construct in the imagination a universe which is in all respects the same as the universe in which we live. And like the scientists of Blueland, Platonic scientists, if successful, will end at the knowledge not only of the actual dispositions and arrangements of our physical universe. They will know also exactly why it is as it is, a knowledge which mere observation of those dispositions and arrangements can never convey.

3. *Timaeus*, 31b.4.

PART II

MATHEMATICS and NATURAL SCIENCE

6

The Divided Line Again

Republic 509b

SOCRATES is pressed by Glaucon to say more of Sun and Good (509 c.5–8). Socrates tells him to take a line, divide it, then divide each of the two sections in the same proportion. This procedure generates a line of four parts, and these four parts are related to each other in the same way as the four kinds of human understanding are related to each other and to the Good. If the parts of the Line are given whole number values according to these proportions; if these four numbers have no factor in common: then the sum of these four numbers is the square number of a whole number. The same number of pebbles used to illustrate the Line may be reformed into a square. This consequence follows whether the divisions of the Line are equal or unequal, but is interesting only where they are unequal.

First a demonstration that this is a consequence of so dividing the Line, and that Plato could have known it to be so. That the consequence holds in particular cases is easily seen. For example, let the Line be divided in the proportions 6:3 and the two sections so formed be divided in the proportions 4:2 and 2:1. Then $4+2+2+1$ equal 9, the square of the whole number 3. Here is an algebraic proof:

> Let there be four whole numbers a, b, c, and d, such that $a/b = c/d = a+b/c+d$ and a, b, c, and d have no factor in common. Then $ad = bc$; $a(c+d) = b(a+b)$. So $ac+bc = b(a+b)$. So, since $a+b \neq 0$, $c = b$. So the hypotheses reduce to $ad = b^2$; a, b, and d have no common factor. Then if p divides a and d, p^2 divides b^2, so p divides b. So a and d have no common factor. Now b divides ad, so

write $b = b_1, b_2$ where b_1 and b_2 are whole numbers and b_1 divides a, b_2 divides d. Then $ad = b_1^2 b_2^2$. So $a+b+c+d = b_1^2 + 2b_1$, $b_2 + b_2^2 = (b_1 + b_2)^2$. But since b_1, b_2 are whole numbers, $b_1 + b_2$ is a whole number. So the sum of the four numbers a, b, c, and d is the square number of a whole number.[1]

This proof is not stated in terms Plato used. But in another form it is like many by Euclid for example. Compare *Elements* 2:4, 8:11, 10:117. Russell and Heath give algebraic paraphrases of these proofs of Euclid which bear a resemblance to the one given above.[2] Heath remarks that the Greeks possessed

.... a geometrical algebra which indeed by Euclid's time (and probably long before) had reached such a stage of development that it could solve the same problems as our algebra so far as they do not involve the manipulation of expressions of a degree higher than the second.[3]

So the theorem and its proof are not obviously beyond Plato's power. Plato himself could have devised the theorem since other theorems about square numbers are traditionally attributed to him. For example, at *Timaeus* 32 Timaeus tells Socrates that between two plane numbers there is only one mean number. Heath takes Plato's plane numbers to mean square numbers and notes that Nicomachus attributed theorem 8:11 in Euclid's *Elements* to Plato. Proclus attributes to Plato a method for discovering sets of three whole numbers which can be the sides of a right-angled triangle. According to Heath, this method was probably derived from a consideration of square numbers and their gnomons represented by dots or pebbles.[4] Equally, contemplation of these figures could generate the theorem proved algebraically above.

The divisions of the Line Plato expresses as a series of ratios in

1. By J. Groves, Dept. of Mathematics, Melbourne University. For $b = c$ see also J. Adam, *The Republic of Plato*, 1921, Cambridge, ii.64.

2. See e.g., B. Russell, *History of Western Philosophy*, 1961, London, p. 54 and footnote: T. Heath, *Euclid's Elements*, 1956, New York, i.3749–381. ii.363f: T. Heath, *Greek Mathematics*, 1960, Oxford, i.91.

3. T. Heath, *Euclid's Elements*, i.372.

4. *Euclid's Elements*, op.cit., ii.294; *Greek Mathematics*, i.89.

the manner of Euclid. These ratios are intended to represent a series of imitations. How? If sets of things replace the magnitudes of the Line, perhaps the smallest set is multiplied the same number of times to produce the second smallest as is the third to produce the fourth and the first and second to produce the third and fourth. If there are x dianoetic images to every one form, then there are x visible images to every one visible object, and x^2 visible images to every one form. This is unlikely to be what Plato meant.

Keats has an image:

And haply the Queen-Moon is on her throne
Cluster'd around by all her starry Fays.[5]

As moon to stars, so queen to fairies. But Keats does not mean strictly to compare the quantities. Moon and stars, queen and fairies are not compared because there is one moon, one queen and as many stars as fairies, but because queen and fairies are disposed in relation to each other as moon to stars, with a gap between the queen and her subjects on all sides as the moon's brightness obscures the stars nearest. As the two dispositions are felt to be similar, so the image is felt to be exact. Such images succeed whatever the relations or ratios between the items of each group, provided only these relations or ratios are parallel. The ratios of the Line do not represent similar numbers or dispositions in space but similar kinds of imitating and imitation, and there are more of them than in Keats' image. There moon and stars are related to each other merely as fairy-queen to fairies, but in Plato's image the intelligible and visible worlds are themselves related to each other as their members among themselves. And so we see how the visible world is so perfect a representation of the intelligible world that it even contains within itself reflections as an image of its own relation to that intelligible world.

My hypothesis is that Plato intended the divisions of the Line to have the mathematical consequences deduced above. The account of the Line is then very much a part of Socrates' account of Justice, comparable to his predecessors' accounts of Justice. For the

---

5. J. Keats, *Ode to a Nightingale*, 36–37.

Pythagoreans Justice was a number equal times equal.[6] The Line is such a number, so Justice is the number of the Line. Justice is as much the coordinating principle of the parts of the Line as of the parts of human state and soul. The quadripartite division of the state into classes and the tripartite division of the soul are complemented by this further division of the modes of cognition into four distinct but related categories. For Parmenides Justice holds Being fast in mighty bonds.[7] In the *Timaeus* the best bond is said to be the accomplishment of proportion.[8] Perhaps too Justice is Plato's binding of being and becoming in his dialogue on Justice.

Where is Justice on the Line? The intelligible sections seem exclusively mathematical. This omission of Justice is odd in a dialogue on Justice, here explaining the knowledge which the ruler must have to rule justly. Plato often claims that to be virtuous one must know virtue. But on the hypothesis Justice is accounted for: not itself located on any part of the Line, it is the structure of the Line and of the universe which the Line illustrates. This Justice the philosopher contemplates, and this contemplation unfits him temporarily for legal disputation about the statues and shadows of Justice in the Cave. Perhaps for Plato anyone who does not see the consequences of his so dividing the Line does not see Justice itself. Obvious from one point of view but not from the Cave, Plato does not choose to explain it.

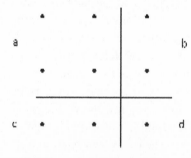

Plato's Line may be represented as a square of dots or pebbles, e.g: where a = 4, b = 2, c = 2, d = 1, the divisions corresponding to the divisions of the Line. The four orders of the intelligible and visible worlds are represented by the four parts of the square in mutual proportion. Let the square a in the diagram stand for the mathematical ideas; b, c and d for the other three orders of

6. Aristotle, *Magna Moralia I*, 1182 a 14.
7. Parmenides, 8, 14.
8. *Timaeus*, 31c 1–4.

being. Then b, c, and d together make a gnomon to the square a. A scholiast to Euclid writes that the name of the gnomon came from its incidental property, namely that from it the whole is known, whether of the whole area, or of the remainder when it is either placed round or taken away.[9]

The point of the remark is that the word 'gnomon' appears to derive from a common Greek verb meaning 'to know', and to be formed by the addition of an ending which denotes an active agent—hence 'knower'. If Plato intended his divisions of the Line to make a gnomon and a square and if he took the derivation of the word 'gnomon' seriously, then he may be illustrating how the three sets of things, b, c, and d, are a means by which the mathematical ideas, a, are known. Or again how the mathematical ideas, together with the material world they generate, reconstitute the overall shape of the mathematical ideas by themselves.

Socrates uses the division of the Line to illustrate a difficult distinction between philosophical and mathematical knowledge. From the postulates of mathematics, the line, the square and so on, the philosopher 'ascends' where the mathematician 'descends.' Could it be that realizing how the divisions of the Line make a square is an instance of philosophical 'ascension'? Clearly it is not the simple apprehension of the theorem that such divisions constitute a square which matters here. That is part of mathematics. But the apprehension that the four modes of human understanding are bound together so closely, and their corresponding objects likewise, this apprehension begins to meet that hardest criticism of Plato's cosmology: why, if the ideas are so wonderful, have a created world at all? Now the orders of being and becoming are spread out in their relations before us and we see that nothing is lost to us in overall perfection by this addition of the creation, and a new intellectual beauty is gained. It is hard usually to accept Plato's distinctions between the more and the less real, but in this passage the 'levels' of reality are so united that we accept their gradations for the pleasure of seeing their harmony.

9. *Scholium* 2, no. 11, Euclid, Editor: Heiberg, v.225; Heath, *Euclid's Elements*, i.371.

Socrates does not explain that the divisions of the Line make a gnomon around the mathematical ideas. This organizing of the world around numbers is familiar to us from the Pythagorean Philolaus, a contemporary of Socrates:

> But in fact Number, fitting all things into the Soul through sense-perception, makes them recognizable and comparable with one another as is provided by the nature of the Gnomon. . . .[10]

This obscure reference to the gnomon is very close in its context to the epistemology of the Divided Line.

10. Philolaus, D–K fr. 11, trans. Freeman.

7

The Nuptial Number

Republic 545d–546d

How, then, Glaucon, I said, will disturbance arise in our city, and how will our helpers and rulers fall out and be at odds with one another and themselves? Shall we, like Homer, invoke the Muses to tell 'how faction first fell upon them,' and say that these goddesses playing with us and teasing us as if we were children address us in lofty, mock serious tragic style?

How?

Somewhat in this fashion. Hard in truth it is for a state thus constituted to he shaken and disturbed, but since for everything that has come into being destruction is appointed, not even such a fabric as this will abide for all time, but it shall surely be dissolved, and this is the manner of its dissolution. Not only for plants that grow from the earth but also for animals that live upon it there is a cycle of bearing and barrenness for soul and body as often as the revolutions of their orbs come full circle, in brief courses for the short-lived and oppositely for the opposite. But the laws of prosperous birth or infertility for your race, the people you have bred to be your rulers will not for all their wisdom ascertain by reasoning combined with observation but they will escape them, and there will be a time when they will beget children out of season. Now for divine begettings there is a period comprehended by a perfect number, and for human by the first in which augmentations dominating and dominated when they have attained to three distances and four limits of the assimilating and the dissimilating, the waxing and the waning, render all things conversable and commensurable with one another. Whereof the basal four thirds wedded to the pempad yields two harmonies at the third augmentation, the one the product of equal factors taken one hundred times, the other of the same length one way but oblong, a hundred

57

of the numbers from the rational diameters of the pempad lacking one in each case, or from the irrational lacking two, and a hundred cubes of the triad. And this entire geometric number is determinative of this thing, of better and inferior births. And when your guardians, missing this, bring together brides and bridegrooms unseasonably, the offspring will not be wellborn or fortunate.[1]

The obvious difficulty with this passage is the number which Socrates propounds. This number is the key to successful marriages and it is often called the nuptial number. Surprisingly, Socrates' complicated number is less difficult to make out than it looks, and there is general agreement among scholars that the number which Socrates has in mind is 12,960,000 or, as we might say, sixty to the fourth power.

The 'number for human begettings' is 36 which is the smallest number with four sets of two factors, the 'four limits', of which three sets are of unequals, the 'three distances'. For 36 = 18 x 2; 12 x 3; 9 x 4; and 6 x 6. 6 x 6 is not a distance (lit. 'apostasy') because its factors are equal, but it is a limit. These four sets of factors may be represented as four plane rectangles. In each of these the longer side, the larger number, is the base. Of the numbers 18, 12, 9 and 6, only 12 is four thirds of another whole number. This 'basal four thirds' multiplied by 5, the pempad, yields 60 which is multiplied by itself three times.

For our purpose we need examine only the last of what Socrates tells us about this number, that it yields two harmonies. The first of these harmonies, we are told, is the product of taking equal factors one hundred times. Let us divide 12,960,000 by 100 and find the square root of the remainder. The square root of 129,600 is 360. So our first harmony is 360 x 360 x 100 = 12,960,000. The second harmony is more complicated. The first factor is common to both harmonies, 100. The second factor is one of the numbers from the rational diameters of a pentad each lacking one, or from the irrational diameters each lacking two. The third factor is one hundred cubes of the triad. The triad is 3; 3 x 3 x 3 = 27; 100 x 27 = 2700. So we have the first and last of the three factors of our second

---

1. Trans., Paul Shorey.

harmony, 100 and 2,700. Let us divide 12,960,000 by both of these factors. The result is 48. This must be the number from the rational diameters of the pentad each lacking one, or from the irrational diameters each lacking two. Socrates makes no distinction here between 50 and the square root of 50.

We have our two harmonies, whatever they are. The first is 360 x 360 x 100, and the second is 100 x 2,700 x 48. Why are we given only these two sets of factors for the nuptial number? 12,960,000 is divisible into many sets of three factors. Why does each of our two sets have three factors rather than two or four or any other number? Let us suppose that what Socrates has in mind are two rectangular solids. Cubes are sometimes called harmonies because they have six sides, eight corners and twelve edges, and 6:8:12 is an harmonic progression. Rectangular solids have the same property. These solids are measured by their three dimensions of length, breadth and height. The multiplication of these three dimensions by each other gives the volume of the solid. We know also that our two sets of three factors have one factor in common, 100. Let us make 100 the height of each of our solids. Then we may imagine the first solid as 360 long, 360 broad and 100 high and the second solid as 2,700 long, 48 broad and 100 high. The first solid is square and squat, a cube cut off more than two thirds of the way down. The second solid is oblong, very long and thin. But they are the same height and have the same volume.

Two of the six faces of the square, squat solid are squares, 360 x 360, and these two squares are its base and its top. The other four faces of the square, squat solid are rectangles, 360 x 100, and these four rectangles are its sides. All six faces of the long, thin solid are rectangular. The two rectangles at its ends are 48 x 100, the two rectangles at its base and top are 48 x 2700; the two rectangles on its long sides are 100 x 2700. Imagine these two solids resting on the ground in front of us. Since this is the nuptial number, let us drive the long, thin solid like a spear through the square, squat solid. We drive one of the two smallest sides of the long, thin solid (48 x 100) through the middle of one of the sides (360 x 100) of the square, squat solid. Our geometrical manipulation was taken to represent a marriage and it is where male and female join which most attracts our eyes when we witness these things in the flesh. We may represent the conjunction of the solids by this plane figure:

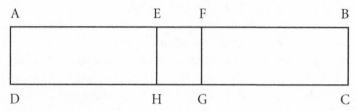

ABCD is one of the four smaller sides of the square, squat solid; EF GH the area cut out of the middle of ABCD by the passing through it at right angles of one of the two smallest sides of the long, thin solid. AB and DC are 360. AD, EH, FG, and BC are 100. EF and HG are 48. AE, DH, FB, and GC are equal.

AB and DC are 360 and 360 is the number of the degrees at the centre and on the circumference of a circle. Let us take it that Socrates intended AB and DC as rectilinear representations of the circumferences of two circles. EF and HG cut out 48 degrees each on the arcs of these two circles. If, standing somewhere on the surface of the earth, we draw a line on the earth through the two poles and through where we stand, we make a great circle of 360 degrees, the centre of which coincides with the centre of the earth. This great circle is called the meridian of a place by astronomers. If now we mark out on this great circle the extreme points north and south which the sun reaches directly overhead in its annual course and we then measure the distance between these two points, we find that it comes to nearly 48 of our 360 degrees. This is the distance in degrees between the tropics of Cancer and Capricorn. Later Greek authors computed the distance between the tropic and equator as equal to the angle subtended at the centre of a regular fifteen sided polygon by each of its sides. This angle is 24 degrees.

Let us now imagine another great circle, this time drawn through the celestial and not the terrestrial poles and passing directly over where we stand. This circle, though much larger than the one drawn on the earth, we may also measure into 360 degrees. If we now mark out on this circle the extreme points north and south which the sun reaches overhead in its annual course and then measure the distance between these two points, it again comes to nearly 48 degrees. So let us take it that in our figure A B represents the celestial meridian, D C the terrestrial meridian, and E F G H the area cut out between heaven and earth by the annual oscillation of the sun north and south, so far as we observe it from most places. This last qualification is necessary because the sun actually cuts out 48 degrees of each circle on both sides of the earth, 96 of the 360 degrees of each meridian. The numbers which Socrates gives us represent the sections of arc cut out by the sun on one side of the earth only, in relation to the two complete circles of the celestial and terrestrial meridians.

This interpretation of the nuptial number may be anachronistic in its assumption that Socrates and Plato divided circles into 360 degrees. We know that this division was not generally applied to circles until later. But the division of the zodiac into 360 degrees, each degree a day of the sun's annual course, was certainly older than Plato. The application of this measure to other astronomical circles, in this case the meridians, is a smaller step than its application to all circles and one we might expect to precede it. There are also in the *Republic* those bonds holding the universe together like the girdles of triremes. Socrates attributes the nuptial number to the Muses and it is the number which his philosopher rulers must know to breed properly. The constitution which Socrates has described in the *Republic* up to this point he believes to be the primordial constitution from which all others have devolved. For all these reasons we must take it that Socrates did not present his number as original with himself but of great antiquity.

Let us now add detail to the plane figure, shading the inner rectangle where the two solids interpenetrate and marking the line between the celestial and terrestrial equators where two doors meet. What this figure suggests is a very good reason for Socrates to think that his harmonies have something to do with human generation. This is as good a picture of the bride on her wedding night as a rectilinear mathematics is likely to provide. It fairly represents the sun god's view of his way to the heart of desire. The figure now illustrates the similarity between the astronomical and physiological formations and we see that on this interpretation Socrates is making analogies between day and night and male and female.

Imagine you are standing at some point on the Equator. The Sun will be directly overhead this point at midday at the Equinoxes. Imagine, too, that you are looking West to the Western horizon, and that you have carefully marked in times past the most southerly point just touched by the Sun's orb as it sinks below the horizon at the Southern Solstice. You have also marked the most northerly point at the Northern Solstice, and the point where the centre of the Sun's orb sinks below the horizon at the Equinoxes. Imagine now that these three points mark the threshold of a great doorway reaching up to Heaven. Beginning from the furthest point South touched by the Sun, draw a line straight up into the sky. This line is the southern doorpost of the doorway. From the farthest point North do the same to make the northern doorpost. The threshold of the doorway is the section of the Western horizon between the terrestrial Tropics; and the lintel overhead is the corresponding arc of the celestial meridian between the celestial Tropics.

Imagine now that this doorway in the Western sky is filled by a pair of great doors, as high as the Heaven. The door on the left has as its doorpost the line drawn straight up from the southernmost point. The door on the right hangs from an equivalent post in the North. The junction formed when the two doors are closed together is the line from the terrestrial Equator to the celestial Equator. We may imagine that during the six months the Sun is south of the Equator, the door on the left is open. During the six months the Sun is north of the Equator, the door on the right is open. At the Equinoxes both doors are open. Or perhaps they are both shut then and through the tiny aperture between them the Sun passes.

We have imagined the doorway as based on the section of the Western horizon between the extreme points of the Sun's southerly and northerly courses. These points are determined by the eye alone. The same is true of the Sun's risings over the Eastern horizon. At sunset and sunrise we can mark most easily the Sun's courses against the earth. But the doorway we have constructed with its doors may be said to stand at every point along the Sun's journey, though the exact location of its doorposts and doors are indeterminable by the eye except at setting and rising. On every meridian, on every line of longitude drawn round the globe through the poles, we may theoretically mark off the section between the Tropics to serve as the threshold, and the corresponding section of the celestial line of longitude to serve as lintel. In this way we may think of the Sun as passing through the doorway at every point on its journey. When the Sun sinks in the West, we think of it as entering through the doorway; when the Sun rises in the East, we think of it as leaving through the doorway. But it is just passing through the doorway at every moment and may equally well be imagined as entering or as leaving or as doing both simultaneously at every instant.

So far we have considered the doorway only in respect of the Sun's daily journey from East to West, by which it completes an entire circuit of the earth every day and night. But the Sun also moves against the fixed stars, relative to which its position changes as they all revolve around the earth. In relation to these fixed stars, the Sun fails to keep pace with them by a little less than one degree a day, and it takes a whole year to complete its retrograde circuit round them all. And while the passage of the Sun through the doorway on its daily circuit is the passage of a point between limits vastly wider than itself, the passage of the Sun through the doorway on its yearly circuit completely spans the distance between the limits of the doorposts twice. And though it is paradoxical it is quite proper to speak of the Sun's yearly journey through the fixed stars as a passage from West to East. If we imagine for a moment the fixed stars actually stopped in the sky but the Sun continuing to move in its usual relation to them, then indeed the Sun would slowly move across the sky from West to East, taking six months in one continuous day before disappearing for six months completely. From this point of view the Sun passes through the doors of the doorway going from West to East, just as Apollo is represented in his chariot emerging from the Eastern pediment of his temple at Delphi, above the doorway and double doors of his temple.

This doorway is a symbol of that doorway we have been cutting into the illimitable sky over the last paragraphs. But it is still some conceptual distance from Apollo's temple at Delphi to the movements of the Sun. Building the Sun's doorway into the walls of a temple at once removes the ambiguity between entering and leaving which was a feature of the doorway by itself. On the other hand the pediment of the emerging Apollo is appreciated typically by someone entering the temple. Leaving the temple, one would be turned away from it. The God's emergence balances the worshipper's entrance. The doorway in the sky was determined entirely by the Sun's movements and the horizon. They were the only material phenomena in an otherwise undifferentiated and limitless expanse. The temple is the solidification of this empty space, which is limited in this representation by the ends of the Eastern wall in which the doorway is set. We may imagine the Sun's journey through space not as the passage of a solid body through a vast, dark void but as the penetration through what is dense and heavy by something extremely light and mobile. In this symbolism the interstellar spaces are assimilated to earth and the passages of Sun, Moon and planets to the tunnels and chambers in a labyrinthine building, say, or a system of underground caves.[2]

In these ways and with these qualifications the doors of Apollo's temple may be compared to the doors of the Sun, and the temple itself may be compared to the cosmos, geocentrically considered. But it is a truism that where temple and cosmos are analogous to each other both will also be analogous to the human body.[3] The cosmos, the temple and the human body are the three primary houses of the spirit. To what in human anatomy do the doors of the temple correspond? The answer is clearly the female *labia majora.* The name Delphi was closely related to the Greek word for womb, *delphys,* and here too was the navel stone, the *omphalos,* which marked the first point of creation.

This account of the doorway began by constructing it on the Western horizon, after marking the Sun's settings. To think of the Sun's journey through the doorway as from East to West is less comprehensive than to think of the movement as from West to East. But at sunset particularly, Venus often appears in great glory on her mount and even the Sun is afraid of her. This Venus is Justice, who measures

2. For example, Plato, *Phaedo,* 111–130.
3. See e.g., Ananda Coomaraswamy's essay, "An Indian Temple: The Kandarya Mahadeo," in *Selected Papers Vol. 1: Traditional Art and Symbolism,* ed. R. Lipsey.

the Sun's courses. No longer the laughter-loving Goddess of Homer, now she is much-punishing[4] and the agents of her justice are the Furies formed at her birth.[5]

The early *Hymn to Pythian Apollo* describes how the god came to Krisa, a foothill facing west beneath snowy Parnassus, and there decided to site his temple.[6] Then he laid out all the foundations, wide and very long. The longer sides of the building were aligned east and west; the shorter sides, in one of which was the main entrance, north and south. The temple was not parallel with the equator but inclined toward it in the west. The eastern pediment showed Apollo in his chariot from the front. Guided by Justice Apollo surveyed the earth from his chariot, going from west to east in his yearly journey round the zodiac. In his temple he was the geographer of the Greeks. His laying out of its foundations symbolizes his nature. The lines he drew on the surfaces of earth and heaven ran straight as the arrows he shot from his bow. With this bow he killed Tityos[7] and also the dragon from whose decay Delphi was called Pytho.[8] These were victories of the mind which measures rather than of the reason which overcomes passion. He is pictured on many vases seated on the omphalos inside the temple, his bow and quiver hanging from a peg. The omphalos is covered with a net or knotted ropes, other symbols of geodesy.

This is Apollo Loxias, Apollo the oblique, a title which describes at once the slanting of the sun's yearly course to the equator and the indirection of his oracles. Apollo Loxias was, most of all, the god of Delphi. He is the subject of Parmenides' poem and the solution to the Muses' number of generation. According to one late author, the Pythagoreans believed that:

> The zodiac runs obliquely on account of the generation of those earthly things which become complements of the universe. For the

4. Parmenides, Frag. L.
5. Heraclitus, Frag. 94.
6. Hesiod, *The Homeric Hymns and Homerica*, op. cit., 287–299.
7. *Odyssey,* XI, 576–581.
8. Hesiod, ibid., 357–358.

passage of the sun and the other planets effects the four seasons of the year which determine the growth of plants and the generation of animals. For if these moved evenly, there would be no change of seasons of any kind.[9]

In *On Coming-to-be and Passing away* Aristotle describes the effects of these heavenly movements in a way which recalls not only Plato's account of the nuptial number but the circular movements of the same and the different in the *Timaeus*:

> It is not, therefore, the primary motion which is the cause of coming to be and passing away, but the motion along the inclined circle; for in this there is both continuity and also double movement, for it is essential, if there is always to be continuous coming to be and passing away, that there should be something always moving, in order that this series of changes may not be broken, and double movement, in order that there may not be only one change occurring. The movement of the whole is the cause of the continuity, and the inclination causes the approach and withdrawal of the moving body; for since the distance is unequal, the movement will be irregular. Therefore, if it generates by approaching and being near, this same body causes destruction by withdrawing and becoming distant, and if by frequently approaching it generates, by frequently withdrawing it destroys; for contraries are the cause of contraries, and natural passing away and coming to be take place in an equal period of time. Therefore the periods, that is the lives, of each kind of living thing have a number and are thereby distinguished; for there is an order for everything, and every life and span is measured by a period, though this is not the same for all, but some are measured by a smaller and some by a greater period; for some the measure is a year, for others a greater or a lesser period.

> The evidence of sense perception clearly agrees with our views; for we see that coming to be occurs when the sun approaches, and passing away when it withdraws, and the two processes take an equal time.[10]

---

9. *The Pythagorean Sourcebook and Library,* op. cit., Photius, 12.

10. Aristotle, Vol. III, *On Coming-To-Be and Passing Away,* Trans: E. S. Forster, 1978, London, Loeb Classical Library, 336a–336b.

8

Numbers of the World Soul

Timaeus 35b–36d

1

There are two series: 1, 2, 4, 8 and 1, 3, 9, 27. Of these numbers 1, 4 and 9 are square numbers. Between two square numbers there is one geometric mean term which is a whole number (*Timaeus* 32a). Between 1 and 4 the geometric mean is 2 ($\frac{1}{2} = \frac{2}{4}$). Between 1 and 9 it is 3 ($\frac{1}{3} = \frac{3}{9}$). 1, 8, and 27 are cube numbers and between two cube numbers there are two geometric mean terms which are whole numbers (32b). Between 1 and 8 the two means are 2 and 4 ($\frac{1}{2} = \frac{2}{4} = \frac{4}{8}$). Between 1 and 27 the two means are 3 and 9 ($\frac{1}{3} = \frac{3}{9} = \frac{9}{27}$). There is also one geometric mean term between the two square numbers 4 and 9. This is 6 ($\frac{4}{6} = \frac{6}{9}$). And there are two geometric means between the two cube numbers 8 and 27. They are 12 and 18 ($\frac{8}{12} = \frac{12}{18} = \frac{18}{27}$).

Fig. 1

```
                 1
             2    3
          4    6    9
       8   12   18   27
```

In this figure there are three rows of four numbers: 1, 2, 4, 8; 1, 3, 9, 27; 8, 12, 18, 27. All three rows of four numbers are in double geometric ratio. There are three rows of three numbers: 4, 6 9; 2, 6, 18; 3, 6, 12. All three rows of three numbers are in geometric ratio. The figure exhausts the geometric and double geometric ratios binding the seven numbers 1, 2, 3, 4, 8, 9, 27. The seven numbers are striking for the number and pattern of these ratios. According to Theon the seven numbers are the second of the several Pythagorean quaterna-ries or sacred forms of number.[1]

If the base of the triangle is 1, 2, 4, 8, then the horizontal rows above that base are 3, 6, 12 and 9, 18. These rows and the base are in duple ratio. If the base of the triangle is 27, 9, 3, 1, then the horizontal rows above that base are 18, 6, 2 and 12, 4. These rows and the base are in inverse triple ratio. If the base of the triangle is 8, 12, 18, 27, then the horizontal rows above that base are 4, 6, 9 and 2, 3. These rows and the base are in the ratio 2:3.

The two series, bound so closely within themselves and to each other, now have their bonds doubled and redoubled by the interpo-sition of their arithmetic and harmonic means as well. Between 1 and 2, 2 and 4, 4 and 8 the arithmetic mean terms are $\frac{3}{2}$, 3, and 6. The harmonic means are $\frac{4}{3}$, $\frac{8}{3}$, and $\frac{16}{3}$. Between 1 and 3, 3 and 9, 9 and 27 the arithmetic means are 2, 6, and 18. The harmonic means are $\frac{3}{2}$, $\frac{9}{2}$, and $\frac{27}{2}$. Let us arrange these numbers in order of increas-ing size, starting from 1 and ending at 27.

This order of numbers is completed by adding new numbers to the point where no two consecutive numbers between 1 and 27 have a ratio to each other greater than 8 to 9. Of the numbers so far, the next smallest to 1 is $\frac{4}{3}$ which is the harmonic mean between 1 and 2. The gap between 1 and $\frac{4}{3}$ can be filled by two times $\frac{9}{8}$ in geometric ratio since 1 x $\frac{9}{8}$ x $\frac{9}{8}$ = $\frac{81}{64}$, which is less than $\frac{4}{3}$:

1, $\frac{9}{8}$, $\frac{81}{64}$, $\frac{4}{3}$, $\frac{3}{2}$...

The third gap here, between $\frac{81}{64}$ and $\frac{4}{3}$, is much smaller than $\frac{9}{8}$. It is $\frac{256}{243}$.

If, on the other hand, the gaps between the means and extreme of the triple series are filled, then the next smallest number to 1 is $\frac{3}{2}$ which is the harmonic mean between 1 and 3. The gap between 1 and $\frac{3}{2}$ can be filled by three times $\frac{9}{8}$ since 1 x $\frac{9}{8}$ x $\frac{9}{8}$ x $\frac{9}{8}$ = $\frac{729}{512}$ which is less than $\frac{3}{2}$:

1, $\frac{9}{8}$, $\frac{81}{64}$, $\frac{729}{512}$, $\frac{3}{2}$ ...

1. *Mathematics Useful for Understanding Plato*, 2, xxxviii.

Here it is the gap between $729/512$ and $\frac{3}{2}$ which is the much smaller $256/243$.

There are, then, two slightly different sets. Do we take the one with $4/3$, or the one with $729/512$, or do we take both? We must take $4/3$ because the harmonic mean between 1 and 2 is specifically mentioned. Do we take $729/512$ as well? The task here is to fill the gaps left in the double and triple series after interposing their arithmetic and harmonic means. The smallest gap so created is not between 1 and $\frac{3}{2}$ but between 1 and $4/3$ since $4/3$ is smaller than $\frac{3}{2}$. This is the only gap to be filled, provided that the two series are taken together as a single order of numbers increasing in size from 1 to 27.

On this basis, the following table represents the seven numbers from 1 to 27 with their arithmetic and harmonic mean terms and with the additional numbers filling the gaps so that none is larger than $9/8$.

Fig. 2

INTERVALS

O	1	$9/8$	$81/64$	$4/3$	$3/2$	$27/16$	$243/128$
C **T**	2	$9/4$	$81/32$	$8/3$	3	$27/8$	$243/64$
A	4	$9/2$	$81/16$	$16/3$	6	$27/4$	$243/32$
V **E**	8	9	$81/8$	$32/3$	12	$27/2$	$243/16$
S	16	18	$81/4$	$64/3$	24	27	

Figure 2 shows all the notes of four consecutive diatonic octaves and the first six notes of the fifth octave. The octave is of the form: tone, tone, semitone, tone, tone, tone, semitone. The table includes all the quantities mentioned in the making of the world-soul (35b–36b). In particular the table includes all the arithmetic and harmonic mean terms in the three triple intervals 1 to 3, 3 to 9, 9 to 27. The sum total of all the quantities of the soul-stuff is $273\,281/384$.

The quantities are added to each other as lengths, end to end, to make a strip $273\,281/384$ units long. Divide this strip lengthwise to make two half-strips, the Circle of the Same and Circle of the Different. Let the Circle of the Same be $273\,15/16$ long. This is longer than $273\,281/384$ by $79/384$. But $273\,15/16 \times 4/3 = 365\frac{1}{4}$.

Let the Circle of the Different be 273⅜ long. This is shorter than 273²⁸¹⁄₃₈₄ by ¹³⁷⁄₃₈₄. It is shorter than the Circle of the Same by ²¹⁶⁄₃₈₄ and fits inside the Circle of the Same (36c). Divide the Circle of the Different lengthwise again, this time into two quarter-strips, each 273⅜ long.

One of these quarter-strips is the outermost Circle of the Different which fits just inside the Circle of the Same. Divide the other quarter-strip exactly into six lengths:

10⅛, 20¼, 30⅜, 40½, 81, 91⅛

These six lengths and the outermost circle of 273⅜ are multiples by 10 ⅛ of 1, 2, 3, 4, 8, 9, 27. In this series there are three double ratios: 1 to 2, 2 to 4, 4 to 8. And there are three triple ratios: 1 to 3, 3 to 9, 9 to 27(36d). The seven circles of the Different are ⅟₃₆, ⅟₁₈, ⅟₁₂, ⅟₉, ²⁄₉, ¼ and ¾ of 364½.

At 273¹⁵⁄₁₆ the Circle of the Same is three quarters of the days in a year of 365¼ days. At 273⅜, as a whole or in sum, the Circles of the Different are seven simple fractions of the days in a year of 364½ days. 273¹⁵⁄₁₆ and 273⅜ do not quite add up to 273²⁸¹⁄₃₈₄ x 2. They fall short by ⁵⁸⁄₃₈₄. This remainder may be the 'seconds and thirds' from which all other living souls are made (41d). Clearly, the measurings of the soul-stuff to make the heavenly circuits are less exact than at first appears. Hence my slight departures on either side of the sum of the musical series. I do not know why the sum total of the notes to the sixth of the fifth octave is a number useful in astronomy. Nor do I know how this choice of the number of days in human gestation links with the nuptial number, if at all.

The seven circles of the Different are the orbits of the seven planets, from the Moon at 10⅛ long to Saturn at 273⅜. The planets correspond to seven tones as well as to the numbers from 1 to 27. 1, 2, 4, and 8 are octaves, 3 is a fifth, 9 a second and 27 a sixth. The smaller number is the higher tone. Of the seven planets the moon's motion is the fastest.[2] The faster motion corresponds to the higher tone.

2. *Republic,* 617a.

Fig. 3

Moon	1	First note
Sun	2	First in second octave
Mercury	3	Fifth in second octave
Venus	4	First in third octave
Mars	8	First in fourth octave
Jupiter	9	Second in fourth octave
Saturn	27	Sixth in fifth octave

The numbers attributed to the planets here do not correspond to their actual relative orbits, periods, speeds, distances from each other, sizes or any other property. Plutarch appreciated the problem and dodges it:[3]

> Just as one is ridiculous, then, who looks for the sesquitertian (⁴⁄₃) and sesquialteran (³⁄₂) and duple ratios in the yoke and the shell and the pegs of the lyre ... while of course these too must have been made proportionate. ...

We too, with Plutarch, will leave aside the *physics* of the celestial concert for fear of seeming absurd. Only the intervals matter which we now understand from first principles, e.g.

Fig. 4

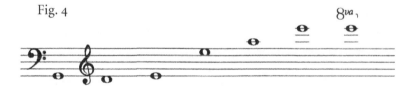

---

3. *Generation of the Soul,* 33.

II

There are seven numbers, seven musical tones and eight astronomical orbits. The seven numbers are two series, the one in duple and the other in triple ratio. This is how Timaeus appears to describe the hoops when the orbits are being made. Another striking feature of the seven numbers is that the last number, 27, is the sum of the preceding six. The seven numbers are often arranged in an inverted V or capital Greek lambda as we have arranged them in the first figure if we remove 6, 12, and 18. This *lambdoma* is one of the enduring figures of Platonism, and it takes conscious effort to remember that the seven numbers are introduced by Timaeus merely as the lattice-work on which to hang the other twenty seven tones in Figure 2.

We should certainly compare the pyramid in figure 1 with the incomplete table or grid of figure 2. The pyramid comprises ten numbers from 1 to 27; the grid comprises thirty four numbers from 1 to 27. In the pyramid all the series of three and four numbers are in geometric and double geometric ratio. In the grid every horizontal line of numbers is a regular geometric series in ⅑ ratio except for the semitones; every vertical line of numbers is a regular geometric series in duple ratio including the semitones.

There is no explicit reference to musical tones in the passage. We are dealing at most only with harmonics. But there is an implicit appeal to harmonics in the ⅑ ratio, the $256/243$ ratio, and above all in the duple ratio of the octave. Quite aside from the interpretation of Plato, there is a very strong argument for supposing that the series of numbers in figure 2 is a natural series where the series of cardinal numbers is to a degree conventional. A decimal or sexagesimal system of counting adopts an arbitrary term, 10 or 60, to facilitate the naming of numbers by repetition. But the turning of the musical scale at the octave is immediately acknowledged by the ear. Again, both the pyramid and the incomplete grid are alive with the energies of their geometrical series. Compare them both to a plain grid of the cardinal numbers to 100 set out in tens to form a square. This square is quite inert except perhaps for the diagonal of multiples of 11.

The grid or table in figure 2 is incomplete since it lacks the last tone in the fifth octave. From this point of view it has been argued that the seven numbers are less satisfactory harmonically than they are as numbers. As numbers they culminate at 27, the cube of the triple series, and 27 is three to the highest power in nature, the solid. But harmonically 27 is merely the sixth tone in the fifth octave. On the other hand, the seven tones are the first in the first octave, the diapason, the double and triple diapasons, the twelfth, the double and triple twelfths. These seven tones are as neat, consistent and simple as one could wish. And the lengths of string under equal tension which produce the first six tones are together the length of string of the seventh tone. In this way, there is a logic to concluding the series at 27. If the seven tones are played consecutively from the highest to the lowest, the series is graceful and elegant and the sixth in the fifth octave a satisfying conclusion.

The thirty four numbers in figure 2 are quantities taken from the mixture by the Demiurge. They are lengths which are added end to end to form a long strip. This strip has a certain width since the Demiurge then cuts it lengthwise. The Demiurge takes the thirty four portions of the mixture in a certain order starting with the seven numbers of the *lambdoma,* then proceeding to their arithmetic and harmonic means in the duple and triple series, and then filling in all gaps larger than ⅑8 with numbers in that ratio or in the ratio of the semitone. But this way of proceeding will not produce the order of numbers in figure 2. Perhaps the order of the numbers in figure 2 is irrelevant. Or perhaps the Demiurge proceeds here like a forensic pathologist with a jumbled skeleton. The bones are picked up randomly, identified, and then placed on a large table, each in its proper place until the skeleton is complete.

Something like this is what the Demiurge must have done, because it certainly matters that the thirty four numbers are in the geometric sequence of Figure 2. The strip produced by adding the numbers to each other will be split to produce the circuit of fixed stars and the circuits of the planets. Those circuits must be as adamantine and indissoluble as they can possibly be. That is why they are manufactured from a geometric series. In this series each term is bound to the one before it and the one after it by a single ratio, ⅑8, or by the semitone. But proportion is the best of bonds. So each part of the line is bound to its neighbors in the best possible way. Of course, a problem arises when the Demiurge binds his strips into hoops. With the largest of these hoops, for the fixed stars and for Saturn, the beginning of the strip, 1, is joined to the end, 27, which is a huge hiatus. But this problem is not resolved by taking the thirty four numbers in the order Timaeus presents them, because then the beginning of the strip, 1, is joined to the last of the numbers before 27, which is 24. And this series makes several other leaps as large, from 27 to 4/3 for example.

The hoop of the Same corresponds to the celestial Equator. The largest hoop of the Different corresponds to the Ecliptic. These hoops are not astronomical lines drawn on a celestial sphere. They are the driving powers of the heavenly movements. The Ecliptic and Equator cross at two points of the Zodiac, the first point of Aries and the first point of Libra, the Spring and Autumn equinoxes. Likewise the hoops of the Same and Different cross at two points: where they are joined together at their centres, and where each strip is joined end to end so as to form a hoop. This second crossing point is a very complex set of joinings. Which crossing point is Aries and which Libra? The point at which the two ends of a strip are joined to make a hoop may reasonably be thought the point at which that hoop begins and ends. Traditionally Aries is taken as the start of the Zodiac and Pisces as its end. So we will take the point at which the strips are joined end to end as the first point of Aries, and the point where the hoops are joined at their centres as the first point of Libra.

As for the angle at which the hoop of the Same crosses the outer-most hoop of the Different at their two junctions, it is 24°, the angle subtended at the centre by one side of a regular fifteen-sided polygon. I have taken this to be the lesson of Plato's account of the nuptial number in the *Republic*. There the angle measures the Sun's course against the equator; here the course of all the planets along the Ecliptic.

The hoop of the Same is a band of determinate length, indeterminate width, and presumably no depth. But it does have an inside and an outside. It corresponds to the celestial Equator and by its revolving in its own space from left to right it drives the whole sphere of the fixed stars, and the seven planets too, in their daily journey round the earth. It circles a largest circumference of the celestial sphere, midway between and at right angles to the Poles. It is hard here not to think of how Parmenides' Goddess Justice holds Being in the limits of mighty bonds. To this hoop of the Same the largest hoop of the Different is joined at two points. This outermost hoop of the Different corresponds to the Ecliptic, and to the course of Saturn around the Ecliptic in an orbit which takes thirty years. As I have shown, this hoop of the Different is only half as wide as the hoop of the Same and it is fastened to the inside of the hoop of the Same.

The outermost hoop of the Different is the orbit of the counter-revolution of Saturn around the Ecliptic from right to left. In this passage Timaeus does not think of the stars and planets as physical bodies but only in terms of their motions. They are orbits, not orbs. Apart from their colors, these orbits were one feature of the stars which the ancients could know with certainty. The stars and planets are Leibnizian subjects which contain all their predicates over time simultaneously. They are created things which we can know as God alone knows us. There is a marked tendency in the early Greek thinkers to think of the heavens as orbits, not bodies, and here the seven orbits of the Different are the framework of the Solar System as the three diapasons and the three twelfths provided the lattice-work of the octaves.

The counterrevolutionary orbit of Saturn is pressed right up against the hoop of the Same. The hoop of the Same drives the whole sphere of the fixed stars, not just the equatorial stars, and Saturn is close to that sphere at all points in its orbit and not just in Aries and Libra. The fixed stars which rise in the East and set in the West were often distinguished from the circumpolar stars which were always visible above the horizon. The fixed stars which rose and set were sometimes conceived as a river called Ocean. So the kingdom of Saturn was along the shore of the River Ocean and this is where Saturn rules over the heroes in the Isles of the Blessed according to Hesiod.[4]

The planetary orbits replicate the diapasons and twelfths in harmonics. The six inner counterrevolutionary orbits together are the

4. *Works and Days*, 167–173.

length of the seventh, just as the sixth tone in the fifth octave required a string the length of the other six together. In this way both the harmonics and the astronomy repeat the numerical containment by 27 of the other six numbers in the *lambdoma*. There can be no question at all of supposing that the planets make actual or notional noises in their passages. The point of Timaeus' disquisition is that the heavens are a visual manifestation of exactly those same principles which the musical scale manifests aurally. These principles are prior to both music and the astronomical cosmos, and are equally realized in both in their quite different ways. Contemplation of these principles in their two different manifestations is the purpose here. It is a way to the integration of seeing and hearing at a point beyond either.

The hoop of the Same is twice as wide as the hoops of the Different, and the hoops of the Different are, taken all together, twice as long as the hoop of the Same. That the hoops of the Different, the counter orbits of the planets, are twice as long as the forward orbit, is the single respect in which the Different here masters the Same. Otherwise it is all the other way and the Same exceeds and dominates the Different to a remarkable degree. Consider the number of counterrevolutions against the number of forward revolutions. In thirty years or 11,000 forward revolutions, Saturn makes 1 counterrevolution; Jupiter 2½ Mars 15; the Sun, Venus and Mercury 30 each, and the Moon some 400. Some 510 counterrevolutions to 11,000 forward revolutions or 1:21.

The imbalance is even clearer when we take into account the distances travelled according to Timaeus by the planets in their counterrevolutions. Since the circuit of the Same is 27 times longer than that of the Moon we now divide the Moon's 400 counterrevolutions by 27 to find the comparative distance travelled. And so with the other planetary counterrevolutions of 2, 3, 4, 8, and 9. This sum shows that during 11,000 forward revolutions, the seven planets together only travel some 31 counterrevolutions of the same length. The motions of the Different may be elaborate but they are minuscule.

These calculations disturb the first sense we have of this passage from the *Timaeus*. Then we felt that the Same and the Different were in balance, that the stuff from which the world-soul was made was made from both equally, and that the hoops of the Same and the Different were likewise quantitatively identical. But in respect of motion the Same is overwhelmingly predominant. This is no Heraclitean coming and going simultaneously in the same place. It is all coming and almost no going. Now of course we can also see or imagine these movements from the Antipodes in which case they will be reversed. Now the diurnal motion will be from right to left and the planetary counterrevolutions from left to right. And there are those Platonic periods when the cycles of the heavens are reversed and we are born out of the grave and disappear at the ends of our lives as babies. Are these two different kinds of reversal enough to restore the balance? Or is the imbalance a symbol of the Divine control?

There is an exact equivalence between the physical bulk of the bands which constitute the Same and the Different, but there is a massive disproportion in the quantity of their motions. But what the Different lacks in quantity of motion it supplies in the variety of motion. For the planets go north and south of the fixed stars; they go faster and slower; and they approach nearer to and go further from the earth. They move in all six directions in relation to the fixed stars and so they are as different as they can be. And this would not have been possible if their counterrevolutions had been much faster than they are. For in that case the planets would have been unable to reverse their countermotions so as to go faster than the fixed stars on occasion, as well as slower. But this is what the five outer planets all do in their retrogradations.

Plato does not discuss all six of these motions in this passage of the *Timaeus*. Here he describes the relative speeds of the planets against the fixed stars in general terms. He describes in detail the movements of the Sun between the Tropics in his account of the nuptial number in the *Republic*. As for the movements of the planets towards and away from the earth, I take this to be what is meant by the varying widths of the lips of the bowls in the Myth of Er.

But there are problems with Plato's scheme. We know, as Plato appears not to know, that the eight hoops of the Same and Different are quite illusory. The heavens do not turn round the earth. They merely seem to do so because the earth rotates once every twenty four hours. The planets, apart from the Moon, do not orbit the earth in their revolutions. The earth orbits the Sun and so do the other planets. Plato has been misled by appearances to construct a cosmology which is based entirely on a naïve geocentric model. We see this most clearly in the *Republic* where the vision of the heavens is preparatory to the descent of the souls through the rings of the fixed stars and planets to a life on this earth. The cosmology is shaped by our earthly destiny. And we must remember that Plato wrote this barely two generations before Aristarchus argued with proofs that the earth revolved around the Sun and rotated around its own axis. Plato is well behind the pace here.

Plato is clearly concerned to provide a description and explanation of how the stars appear to us as we look at them from the earth. He is concerned with how we actually experience the heavens through our eyes, not with the counterintuitive theories of the physicists. He is a theoretician of appearances. Is the world at base a material or a psychological creation? In this passage of the *Timaeus* the psychology precedes the physics which merely embody the revolutions and counterrevolutions of the world-soul in its own self-sufficient contemplation. For Plato the world is at base psychological, it is primarily an instrument for the evolution and devolution of souls as moral and spiritual beings. To that end, the world as given in experience is what matters, how it actually strikes upon our senses. For it teaches us so. Plato is concerned to 'save the appearances'. We, on the other hand, have forsaken the evidence of our own eyes almost completely in this matter in recent times. The abstractions of modern astrophysics seem to have led us to the point where the simplest features of the Sun, Moon and the planets as seen from the earth are unknown. This ignorance is widespread in all walks of life.

Plato gives us an astronomy of these appearances. They are quite steady and constant. They do not change over time. The cosmos which Plato describes in the Myth of Er and in this passage of the *Timaeus* is the same one I inhabit twenty four centuries later on the other side of the globe, when I step outside of an evening. In this way Plato's astronomy is much plainer than the science of our time which acts like the car roof and the electric lamp to mask our heavens from us. We are very late arrivals in the cosmos, in the modern view, and in no way was it made for us. All the night visions of all the people who have ever lived, all the creatures' eyes that have ever looked upwards, do not count much against the weight of our new truths. Ironically, Plato too, though a geocentrist, was opposed to idle star-gazing. The task was to discover the laws which governed the phenomena. These laws were not the mechanical laws of necessity but a mathematics of beauty.

In this theory of astronomical appearances it remains only to show why Timaeus supposes the lengths of the seven planetary orbits to be in the ratios of the *lambdoma*. Distance equals speed times time, but the only member of this trio which the Greeks knew of the planets were the times they took to complete their counter-revolutions. They might assume that the planets all travelled at equal speeds. In that case Saturn's orbit would be nearly 400 times greater than that of the Moon, since the moon completes nearly 400 counterrevolutions in the time that Saturn completes one. But there is no compelling nor even suggestive argument that the planets do all travel at the same speed overall in their counterrevolutions.

Plato did not assume this. He assumed instead that the lengths of the orbits were in the ratios of the *lambdoma*. His assumption was quite in accord with the details of the planetary motions which they knew, the periods of their counterrevolutions. On this assumption the actual speeds of the planets are very varied. The Moon moves more than six times faster than the Sun; Mars moves five times faster than Jupiter; the moon is now fifteen times faster than Saturn. Jupiter is now marginally slower than Saturn, actually though not apparently. All these are quite possible as actual speeds, but it was not mere possibility which led to this assumption of the ratios in the *lambdoma* for the planetary motions. These ratios are numerically and harmonically pleasing and their application astronomically realizes the same principles in space. That is why the orbits have these lengths, the best possible arrangement within the range available to the Demiurge. And the best arrangement among the possibilities available to the earthly astronomer, given the known periods of the counterrevolutions.

After the numbers, the harmonics and the astronomy we come at last to the real subject of Plato's account, the psychology of the world-soul. This is not human psychology. Somehow we must convert our previous scientific speculations into the mentality of the soul or spirit which governs the cosmos. Usually this is achieved through drawing a parallel between the operations of that mind and the movements of the stars. So Plotinus argued that the world-soul desired God but could not be God. For if the world-soul were God, it would no longer be soul. Instead:

> The soul exists in revolution around God to whom it clings in love, holding itself to the utmost of its power near to Him as the Being on which all depends; and since it cannot coincide with God it circles about Him.[5]

The energy of a star's desire for God is converted into the motion of its orbit around a centre which it worships. This is lovely but it seems to suggest that the earth at the centre is the God whom the stars worship, rather than the sublunary earth of change and decay. In the *Paradiso* Dante follows Plotinus in his account of 'the love which moves the Sun and the other stars.' But he also recognized the difficulty that the stars circle round the earth. He ascends through the rings of the planets, in each of which he meets a certain kind of saints and patriarchs. Then he looks back through all the rings and sees the earth at their centre:

> ...and I beheld this globe
> Such that I smiled at its ignoble semblance.
> And that opinion I approve the best
> Which holds it least....[6]

Soon after, an astonishing transformation occurs, the most amazing in a book of astonishing transformations. The rings of the planets and fixed stars suddenly invert themselves, so that the largest and outermost rings become the shortest and most inward around the throne of God. They are, in fact, eight of the nine angelic orders, as well as the homes of the blessed who have lived on earth.

5. *The Enneads*, II, 2.
6. *Paradiso*, 22.

Dante has personalized the planets as the homes of the blessed souls. So Saturn is the place of the contemplatives, the intellectual saints. We may compare this with Hesiod's account of Saturn's kingdom where the heroes live after death, far from the other immortals along the shores of Ocean stream. Plato, too, makes the planetary realms a place for human beings, during life and after death, but much more variously than Dante and Hesiod do.

In the *Phaedrus* Plato describes how the Gods in their chariots rise up at their appointed times 'to the summit of the arch which supports the heavens' where they transcend the physical cosmos and go to their feasting. There are human souls in the retinues of these Gods and Goddesses, and at the time of their divinity's ascent to the summit, those souls do their best to follow. And if the chariot-horses of those souls are properly disciplined and their wings are fully grown, then they too achieve that same transcendence so far as they can. The summit of the heavenly arch is at the equinoxes of each planet's course, at which time the planet is passing over the Equator. In the *Timaeus* these times and places are the intersections of the Equator and the Ecliptical counterrevolution of the planet. At those moments according to the symbolism of Parmenides' chariot-ride, both the doors of the palace of Night are open. They are both open for the few hours during which the planet passes exactly between them. In the symbolism of the *Timaeus* these two points at the Equator and the Ecliptic are the two points where the outmost circle of the Different is attached to the inside of the circle of the Same. Where the Ecliptic crosses the Equator, there time and Eternity are momentarily at one. Of course, only Saturn of the seven planets actually passes through the points of juncture with the Equator on Plato's scheme. The other lower planets have their arches and their summits lower than Saturn's, as they pass through them during their various years and with their retinues.

In the *Phaedo* too, human souls accompany the planets in their spiralling courses. But now these courses are the Rivers of Hell, burning Pyriphlegethon and Cocytus and Acheron, and the souls in them are grievous sinners undergoing their punishments. But even they periodically have a chance of reprieve as the Rivers carry them past their victims in the Acherusian Lake, when they can cry out for forgiveness. The picture of the planets here and of Ocean stream is lurid and ghastly, fit for the *Inferno* and not Paradise.

What Plotinus, Dante, the *Phaedrus* and *Phaedo* all have in common in their psychologies of the world-soul is their use of the heavenly orbits rather than of the musical tones or the seven numbers of the *lambdoma*. Dante makes his planetary realms of saints and patriarchs musical as well as bright, and there is an obvious parallel with the whirling orbit of a bull-roarer and the continuous noise produced. For Timaeus the planetary orbits all begin at once and when eventually the planets complete their revolutions at exactly the same moment, then a great year is concluded. To the other forms of completeness we may also perhaps add this, the exhaustion of all possible relations between the seven orbits. There is a very exact aural equivalent to this method of exhaustion in English bell-ringing, where an octave of bells, let us say, rings out the changes by varying the order in which the eight bells ring until that particular system of changes has completed all the possibilities. This may be a parallel between the visual and musical orders.

The seven numbers of the *lambdoma* are the second Pythagorean *tetractys*. These numbers inform the musical and astronomical manifestations. In the seven numbers we see where these two realms actually coincide. And we feel the sense of the Pythagorean oath:

By Him who handed on to our generation the *tetractys,* fountain of everflowing nature.

9

The Human Head and Face

Timaeus 44d

First, then, the gods, imitating the spherical shape of the universe, enclosed the two divine courses in a spherical body, that, namely which we now term the head.

IN THE *Timaeus* the gods who make the human head are the stars and planets, to whom the Creator gave the task of creating mortal creatures. The Creator could not accomplish this task himself since anything which the Creator made would last forever. The Creator made the Heavens and the Heavens made us.

The human head is the prime creation in the mortal realm because it most closely follows the pattern of the cosmos. For the purposes of Plato's analysis, the pattern of the cosmos has been founded on two related but distinct phenomena, geocentrically considered: the rotation of the fixed stars and the varying rotations of the planets. Plato calls the rotation of the fixed stars the course or circuit of the Same, and the rotations of the planets, all taken together, the circuit of the Different (35). Clearly the rotation of the fixed stars, including the circumpolar stars, generates a sphere, and this is the model or pattern of the human head.

But Plato supposes that both the divine circuits are found in the human head, not just the circuit of the Same. Somehow that other circuit which comprises all the motions of the planets in contradistinction to the motion of the fixed stars, somehow this circuit too is bound into the human head. Plato transposes the cosmic motions into a metaphor of human thinking. When the motions of our thinking correspond to the heavenly motions, then we can see what is the same and what is different accurately. But when the motions of our thinking are disrupted, then we lose the capacity for rational judgement (89e-90d). So disrupted can the interior motions become that they are the very reverse of the proper motions. Then everything appears to the judgement as if it were upside down and left to right. This is what the flood of experiences does to us while we are growing up, and it is only with maturity that the real circuits of the soul can reassert themselves and return us to a harmonious and rational state of mind. In animals this disruption of the proper circuits is not temporary but permanent, and this is reflected in elongations and other distortions in the sphericity of their skulls:

> The race of wild pedestrian animals, again, came from those who had no philosophy in any of their thoughts, and never considered at all about the nature of the heavens, because they had ceased to use the courses of the head. In consequence of these habits of theirs they had their front legs and their heads resting upon the earth to which they were drawn by affinity, and the crowns of theirs heads were elongated and of all sorts of shapes, into which the courses of their souls were crushed by reason of disuse (91e).

Leaving aside comparative cranial morphology, I find it hard to think of my mind or thoughts as moving like the fixed stars and planets in their courses. Plato makes clear that this is a case of 'use it or lose it.' Human beings who do not practise theoretical astronomy, who do not consciously exercise their spherical mental motions, will be reborn as quadrupeds. Physical astronomers who gape at the actual stars without working out their motions from first principles are reborn as birds (91d). Their heads are still round but on thickish necks and rather small, we are left to infer, and they have wings. But they are also, of course, still bipeds.

Plato here is a philosophical Aristophanes. But the point remains that there is little serious evidence for assimilating our minds and thoughts to the cosmic motions. Though the rotation of the fixed stars provides the model of sphericity after which the human head is made, we have not discovered any link between that head and the circuits of the planets. But there is one feature of the human head which may matter here. In the Myth of Er Plato represents the planetary motions as like the rims of several bowls nesting inside each other.[1] These rims form a kind of plane surface like the whorl of a spindle, through the centre of which a shaft or pole passes. The same idea is conveyed by the spinning surface of Homer's Charybdis.[2] The crux is that the motions of the planets, including Sun and Moon, are all unidirectional and they all move within a narrow belt between Tropics like those of the Sun. The planets rarely move more than 26 degrees north or south of the Equator. From this point of view, the Earth stands in relation to the planets as Saturn to its rings.

If we apply this schema to the human head, we draw a band around the middle of the head, the band's width corresponding proportionately to the tropic band around the earth. This band, it seems to me, would cover the eyes and ears like a blindfold. So the locations of the organs of seeing and hearing in the human head correspond to the location of the planetary tropics in relation to the whole earth. Seeing and hearing are, for Plato, the best of all our senses because they are connected to the heavens. Sight has been given us to see the heavens, whose motions stimulate us to conceive of numbers.[3] From the study of numbers comes philosophy, the best gift of God to man. As for hearing it is made to hear the voice. The voice provides rational speech and singing. Singing employs those intervals which organize the planetary motions, so that listening to music helps to re-establish the planetary motions in our own heads. In this way, though music is not connected directly to the heavens, it is organized by the very same principles which organize the planetary motions, and so is cognate with those motions. There is then, a certain aptness in the placing of these two senses within the area cut out by the motions of the planets. They are the divine senses, the binding of the divine circuit of the Different in our heads.

Select any point on the surface of a sphere. Mark the point diametrically opposite to it on the other side of the sphere. Through

1. *Republic*, 616c–617b.
2. *Odyssey*, 12. On this comparison and its astrological symbolism see Roger Sworder, *Homer on Immortality* in *Science and Religion in Archaic Greece* (San Rafael, CA: Sophia Perennis, 2008), pp. 45–49.
3. *Epinomis*, 978; *Timaeus*, 47a–b.

these two points draw two great circles round the sphere at right angles to each other. The sphere now looks like an orange which has been cut into four equal segments and reassembled. Now draw a third great circle equidistant from the two points at which the first two circles crossed and at right angles to those circles. We now have three interlocking circles, each of which is at right angles to the other two, so this figure has much in common with the three dimensional cross.

Applied to the cosmos, the selected point is the celestial North Pole; the point opposite is the celestial South Pole; the first two great circles are two celestial meridians at right angles to each other; the third great circle is the celestial Equator. The same account holds, *mutatis mutandis*, if we apply our sphere to the earth. Applied to the head, the selected point is the fontanelle; the point opposite is the opening to the throat; of the first two great circles, one passes through the fontanelle and the ears, while the other passes through the fontanelle and along the nose; the third great circle passes through the eyes and ears. Of course on this view, strictly, we should have one eye at the front of our heads and one at the back. But, as Plato points out, the gods who made us felt that we needed a distinct forward direction and for this reason shifted all sight to the front and made the human face (45a).

The face is the epitome of the head. The head in all its aspects is more than our seeing can grasp at once. We do not see in the round but from two points on its circumference. The face represents the beauty of the head as seen from the limitation of a narrow point of view. The Sun, Moon and planets shine from our eyes. The eyes move freely in contrast to the face as the planets move variously against the background of the fixed stars. The hairline and the line between the closed lips suggest the arctic and antarctic circles. The point of the chin is a projection of the head's South Pole, to sustain the face's presentation of the head in its entirety. Even the furrows of a worried brow suggest an astronomer's lines of latitude round the Northern Hemisphere. When we look into the face of another person, we are seeing the cosmos as a mortal animal like ourselves. The cosmos, too, is a living animal but an everlasting one, and much too big and too complex for us to see with our eyes and know.

PART III

CRITIQUE of NATURAL SCIENCE
and PHILOSOPHY

10

Natural Science

IS IT TRUE that the head, the face and the human reproductive system are homologous with the movements of the stars and planets, viewed geocentrically? These are questions in morphology, of the same kind as questions concerning the relations between comparable organs in animals or in plants. In our times morphological studies of animals often proceed on the hypothesis that animals evolve from each other, while Plato supposed that they all descended from the human, becoming increasingly many-legged until they form into snakes and eventually disappear into the sea as fish.[1] Compared to seeing the differences between kinds of animals, the shift from studying the stars to seeing comparable formations in the human is enormous. But there is a certain intuitive rightness in Plato's notion that we are in this way microcosmic representations of the whole, and the physical similarities between the Sun's doors and the female doors, between the human face and the planetary system, are immediately engaging, even if *outré*. Many, no doubt, would argue that these macrocosmic analogies to the human are bad science, but they would not, I think, dispute that they were scientific hypotheses. Certainly they are not religious or spiritual claims. Plato may have fantasized as to how these correspondences occur in a creation myth, but even in the myth it is the stars and planets which determine the human form. Even here there is nothing in play beyond what is strictly observable.

From a certain point of view, it is quite surprising that these ways of understanding the human head and reproductive system are not

1. *Timaeus,* 91.

much better known, in outline if not in detail. After all, they make no claims on faith, they analyze observable data according to known methods, and they provide explanations of phenomena which are peculiarly dear to us. Furthermore, these ways of reading ourselves are enshrined in traditional forms still in vogue among us. But the only explanations current concerning the human complex seem to be socio-biological or neo-Darwinist. Reasons for this oversight are not hard to find. Part of our pride in ourselves comes from our having outstripped our ancestors in our knowledge of the cosmos. Our competitiveness insists that the Einsteinian theory superseded the Newtonian, and the Newtonian superseded the Ptolemaic. We would do much better to suppose that the Ptolemaic system provides us with an adequate account of the universe *geocentrically considered*; the Newtonian provides the heliocentric account; and the Einsteinian the relativist account. We need all these accounts but the geocentric most of all, because this is the one which tells us how we ourselves are shaped.

But even if we had retained a lively picture of the geocentric cosmos, would we, then, have infallibly identified the forms of that cosmos with those of the human body? Not, I think, if we were empiricists of the Enlightenment. Even though all the data necessary to these identifications are empirical, and the method is a form of morphological analysis, even so modern empiricism could not entertain these identifications. In the first place they are not quantifiable in the requisite manner. The determination of the identity between ratios in two different dimensions, the astronomical and the zoological in this case, is not a matter of measurement. No imaginable meter could ever compute it. Secondly, the modern empiricist will argue that no physical explanation has yet been offered of how exactly the movements of the geocentric cosmos come to form the human body after their own pattern. No evidence has been advanced even to indicate such a causal link. The identifications turn, in fact, upon an absurd and antiquated aetiology by which cause is to effect as original to image. The human body images the stars and planets. But in modern empiricism cause and effect are merely events bound to each other more or less invariably.

These objections have some weight but they are not conclusive.

There is nothing unscientific about the notion that certain physical features of the human being parallel features of the planets and stars viewed geocentrically. We can easily imagine data which would confirm or contradict such an hypothesis. If we were to find life forms elsewhere in space which were more or less identical to the human on planets unlike our own macrocosmically, or if we were to find life on planets like our own but no life forms like ours in these respects, either of these discoveries would contradict the hypotheses developed here. But until that time we must, I think, admit most of these claims into the ranks of scientifically sound empirical hypotheses.

A low rank, but at least it is respectable by the stern standards of modern empiricism. For the ancients, of course, these identifications of cosmic with human formations were much more than mere so-far-untestable possibilities. They were the observable facts which grounded their understanding. It is just here that we can examine more closely the differences between our empiricism and theirs. The ancients admitted many more analogies than we do and founded their understanding on them. Neither we nor the ancients have moved beyond the observable data. So how are we to judge between these two empiricisms?

The nuptial number demonstrates an identity between astronomical and physiological formations. It is both mathematical and a conceivable hypothesis in natural science according to our own scientific lights. The numbers of the world-soul demonstrate an identity between numerical, harmonic, astronomical and psychological formations. Though these formations are mathematical like the nuptial number, the identity which they establish is rather less amenable to our own scientific thinking. But it is also true that the numbers of the world-soul present a geocentric planetary system in general, where the nuptial number presents the special instance of the Sun's movements within this.

There is a passage in the *Epinomis* which describes very well the multidimensional or multidisciplinary character of Platonic science. The *Epinomis* may or may not have been composed by Plato

but it was surely composed by someone who knew the numbers of
the world-soul in Plato's *Timaeus*:

> To the student who studies in the proper way, every diagram,
> every system of number, every harmonic conjunction, in agree-
> ment with the circuit of the stars, must disclose themselves as one.
> And they will disclose themselves so to the student who looks to
> the one in the way I have described. For one bond, grown of all
> these, will disclose itself to those who ponder.[2]

This is a very lofty scientific aspiration and it fits very well the kind
of thinking which we have studied. And yet it is not only the philosophical critic who may pause here. The prose stylist too may well
object to that threefold use of the word 'disclose' as one affirmation
too many. And this criticism leads naturally to the much more serious complaint that Plato is always pressing too hard to find such
unities as these. His theories, in fact, suffer from the classic failing of
rationalism. He presumes that the world conforms to his precon-
ceptions, fails to experiment or search for counter instances and
collapses at the first breath of decent scientific procedure. And it is
all there in that reiteration of 'must disclose', 'will disclose'.

And our critic may continue that there is not much at all in
Plato's grand theory of the multidimensionality of the *lambdoma* in
numbers, harmonics, astronomy and psychology. For what finally
does it come down to? The seven numbers, it is true, have a certain
toy-like attraction, but after that it is all downhill. When applied to
harmonics the seven numbers leave us with a truncated fifth octave.
When applied to astronomy, they supply an entirely putative series
of lengths of orbits for the seven planets. The best that can be said
for this scheme is that it did not directly contradict anything of
what was known by these sciences at the time. But even from this
point of view it is feeble enough. But because it was all wrapped up
in an almost impenetrable exposition it has not only survived, it has
actually hypnotized the minds of scholars for centuries to their and
our greater confusion.

These criticisms overlook the signal achievement of Plato's the-

2. *Epinomis*, 992.

ory. Plato does two things here. He shows how the harmonic and astronomical orders express a single numerical formula, the *lambdoma*. And he adopts with the *lambdoma* a numerical order which has its own intrinsic justification as a beautiful series of ratios. There are implicit two levels of explanation which work together. There is no question here of Douglas Adams' 42, that inexplicable number which turns out to be at the scientific root of all phenomena. With Plato's formula we can go the next step and explain the formula. The truncated fifth octave is less important than the way in which Timaeus constructs the series of thirty four tones around and upon the *lambdoma,* on three octaves and three twelfths. He does not impose the seven numbers on the series since they are the first numbers of the series to be established. There may be other, easier ways of generating the thirty four tones, but I do not know of any. As for the orbits, any account of their lengths would have to be putative. Plato's account is impressive here for its extraordinary economy. Could there be a simpler way of organizing the eight orbits of the planets and fixed stars as seen from the earth?

The reiterated 'disclosures' of the *Epinomis* passage suggested a certain anxiety. And so, in my view, does the more extreme modern criticism of the *Timaeus*, by George Sarton for example. That modern anxiety may spring from the fact that our own researches into the physical nature of things are forever in danger of ceasing to be science at all. Plato's Socrates distinguished very carefully between the useful and practical applications of mathematics and the philosophical approach to numbers. No general can operate safely without the capacity to compute large numbers, but this facility has nothing to do with the contemplative sciences.[3] In the same way research in a laboratory may require a high level of mathematics but if that research is devoted to any practical end, then it is not scientific research, strictly, but technological research. That most of what is called scientific research is of this second kind has led to a crisis in science. This crisis was already acute in the late eighteenth century and has become more extreme with every decade since. Unlike Plato we have no systematic metascience by which to integrate our

3. *Republic,* 522d.

various physical accounts. As a result they have grown in all directions without system.

Their results have been of the greatest utility but they are not science as they stand. Technological results may be of value to the scientist and scientific results to the technologist but there is no necessary connection. Science is comprehension, not utility. It is undertaken for its own sake and it seeks only to know. For this is a much greater goal then mere use. Nor need the endless proliferation of means assist us one inch further on our way to understanding. William Blake compared the natural sciences of his day to an endlessly enlarging labyrinth. What was missing then, and still more now, is a moratorium on all further research, and a long reconsideration of what we have already discovered. This has only to be suggested for its impossibility to be apparent. And that impossibility illustrates just how far we are from wanting to know and not merely to use.

The mere discovery of the formulae which coordinate different disciplines is not of itself enough, however far reaching the formulae may be. For then we are left merely with some form of 42. But with the *lambdoma* we can give an account of why these numbers are intrinsically satisfying and complete. The numbers of the world-soul would satisfy even that highly critical young Socrates whom Socrates described in his last hours in the *Phaedo*. The young Socrates wanted to know all about the physical cosmos but would not be satisfied with any explanations which did not also show why the world was best arranged as it was arranged. The beauty of the *lambdoma* would meet that requirement. And in this way the numbers of the world-soul are superior to the nuptial number. However significant and exact the formula which coordinates the astronomical and reproductive realms, that formula does not itself possess any particular value. Neither 12,960,000 nor the two rectangular solids. To be sure, the two solids, light and dark, are derived from the one number and to this extent the advice of the *Epinomis* is followed about always looking to the one. But that is hardly comparable to the overarching splendors of the *lambdoma* or the square and gnomon.

This same difference between modes of scientific explanation is very well illustrated by Newton's two approaches to his laws of gravity. Having discovered and established the mathematical formulae which coordinated the heavens and the tides, he wanted to know why these were the governing laws. And he believed he found the answer in supposing that these laws also governed the tuning of musical instruments. From here he could conceive of gravity as a system of attunements, and of the heavens as Apollo's lyre or as the pipes of Pan. Note how easily Newton was pleased.[4] But the numbers of the world-soul go much further than merely establishing the unity between harmonics and astronomy. For they are themselves significant.

This, then, is the achievement of this passage in the *Timaeus*. It is an example of a certain kind of scientific explanation, each part and stage of which is unusually responsive to the question 'why?' As we meet this question successively, we are finally left with explaining the beauties of the *lambdoma* as a tetractys. If only we could also explain just why this figure, of all the beautiful figures and series in mathematics, was chosen to make the cosmos. Still, we get further with Timaeus than with most. And this all raises the question of our own scientific, as opposed to technological, achievements. How effective in comparison are our own modes of explanation? Do our explanations answer as many 'why' questions as the *Timaeus*? And there is the other question: how far is our science concerned to teach and enlighten in any case?

The *lambdoma* is like the square into which the Divided Line may be reorganized. The first is a beautiful numerical series, the second a beautiful geometric plane figure. Series and square are the ultimate principles of organization in these Platonic accounts. This beauty also shows in the simplicity, reach and economy of the explanations which they provide, but this is a beauty found in many theories, ancient and modern. In the science of Plato, the capacity to appreciate *lambdoma* and square for their mathematical beauty alone is vital. But this is not a capacity much developed in those who employ mathematics for practical purposes. They are not trained in this aesthetic, for all that they master Euclid very early in their educations. And this, no doubt, is why Socrates insists upon that harsh distinction between the two uses of mathematics. Those who think

4. Isaac Newton, *Classical Scholia* to *Principia Mathematica*.

of mathematics as a tool are not easily brought to think the science divine. They have not been trained to contemplate the mathematical elements.

That science is primarily a mode of contemplation is a thesis which has been under attack from the beginning of the seventeenth century. Francis Bacon condemned the ancients for contemplating nature when they should have turned their minds to the practical, to the relief of man's estate. This view has prevailed and as a result our researches are in some disarray. In Germany the larger questions concerning the place and status of the mathematical and natural sciences were very keenly discussed. Leibniz and the von Humboldt brothers were the organizers and inspiration of great Academies. For Novalis, Fichte and Goethe the overall shape and direction of the sciences were crucial concerns. Much of this is reflected in the prose works on scientific method by Samuel Taylor Coleridge. Coleridge had met Karl Wilhelm von Humboldt, and had recited Wordsworth's *Intimations Ode* to him. But Coleridge took from Germany a sophisticated understanding of scientific method. He distinguished between the contemplation of reason on the one hand, and mere abstract understanding on the other. Abstract understanding is Coleridge's name for the sciences as generally practised. He describes the contemplation of reason as follows:

> The contemplation of reason ... is that intuition of things which arises when we possess ourselves as one with the whole, which is substantial knowledge.... It is an eternal and infinite self-rejoicing, self-loving, with a joy unfathomable, with a love all comprehensive. It is Absolute....[5]

This sounds much more like the ascent from the Cave to the vision of the Good than a Baconian programme. It sounds even more like contemplating the idea of the beautiful as Socrates reports Diotima in the *Symposium*.

What did Plato himself suppose that he had achieved with his theory of the world-soul? Was it, as Burnet and Taylor supposed, an historical reconstruction of Pythagorean science a generation

---

5. S. T. Coleridge, "Essay on Method" from *The Friend*, 1809.

before Plato's own time? Was it Plato's best guess at the natural order at the time of writing? Or did Plato believe it to be much more than either of these, an account never to be superseded, the last word on the foundations of natural science? It is quite clear that Plato supposed research and development essential to the health of the sciences, but it is just as clear that he supposed the ancient Egyptian civilization much wiser than the Greek. The natural science of the *Timaeus* is 'a likely story' not because it is provisional, but because the certainty which attaches to any account of the created order is far inferior to that of the mathematical sciences. There can never be an account of the created order which is anything more than 'a likely story' because the world of change is radically resistant to a clearer or more certain explanation. But whether or not Plato himself believed that the *Timaeus* accomplished all that could be accomplished in the natural sciences, this is how his work was understood by Platonists and Christians.

And how are we now to value Plato's theory of the world-soul? As the last word on the foundations of natural science from a geocentric perspective? It has been said of the psychology of our time that there is no defence for its inadequacies by claiming that it is a young science. Mathematics was a young science once but all the results of early mathematics are results in mathematics still. Euclid is still a mathematical textbook, and Plato's proofs concerning squares remain valid. Is the *Timaeus* still valid in this way as natural science and psychology? It is enough to have reached the point where the question is asked.

So much for the numbers of the world-soul. With the nuptial number, on the other hand, we have a fine example of long-standing early mathematics. The mathematical determination of the distance between the Tropics at 48° remains to this day, with greater precision achieved by Eratosthenes and others. This measurement, with the scheme of Equator, Ecliptic and polar circles is the foundation of mathematical geography and astronomy. In our epoch we owe its appearance to the Greeks and Plato. As the *Romaka Siddhanta* shows, the Hindus learnt it from the Greeks and Romans, while the Chinese organized their heaven exclusively round the Pole Star.

11

Philosophy of Science

THE SQUARE and the gnomon or, as we would say, the equation $(a+b)^2 = a^2 + 2ab + b^2$, preside over Plato's epistemology in the *Republic*. This pattern has the same role that the *lambdoma* has in Timaeus' account of the world-soul. And as with the *lambdoma* in the *Timaeus*, so in the *Republic* we may feel that the pattern of the square has led Plato to top and tail his material to fit it. How far do these four modes of cognition actually bear the complex of imitations and interrelations he ascribes to them? The noetic mathematician grasps the unhypothesized beginning from which all else flows. This is the vision of the Good which I have identified with Parmenides' revelation of being. But it is so far outside common experience as to be beyond discussion. The dianoetic mathematician practises an established science and proceeds to conclusions by means of visible diagrams. This is recognisable. *Pistis* is the understanding of the world of change in the light of the perfect, intelligible living creature. This is a unity of the natural kinds to copy that unity of the mathematical kinds which is the Good. But it is hardly more recognisable than the object of the noetic mathematician which it parallels. *Eikasia* is the belief that the physical world is all that there is and the only knowledge is knowledge of the world of the senses. The order in which these four kinds of understanding are actually experienced is *eikasia, dianoia, noesis,* and *pistis*.

It is easy to see here how *pistis* has been tailored to fit it for its place in the complex. But the perfect intelligible living creature is a poor match for the unity of the mathematical ideas which is the Good and the unhypothesized beginning. Does this intelligible creature contain the elements and the regular solids? If it does, it is hard to see how it is a unity. If it does not, it lacks the integrity of the Good. In fact, two out of the four modes of cognition related here, *noesis* and *pistis*, are quite outside the general understanding. What of *dianoia* and *eikasia*? How alike are they? The mathematicians use visible diagrams according to Socrates, but they are not discussing or thinking about the diagrams, but the ideas. But they do not know what these ideas are, and take no trouble to find out. The prisoners in the Cave, however, are convinced that the images on the wall are all that there is and will kill anyone who denies it. They have no notion at all that there is anything beyond their experience, except madness. So in this crucial respect the parallel between *dianoia* and *eikasia* breaks down. And this is one of the few places where we might hope to understand and maintain it.

There are more general objections. Noetic mathematicians and pistic natural scientists are very rare if they exist at all, and even dianoetic mathematicians are a very tiny portion of humanity. Are we to suppose that these three tiny subsets are somehow to be categorized as three categories, and the whole of the rest of the human race merely as one category? Certainly Socrates is discussing modes of cognition here, not divisions of humankind. But Plato himself ridicules those who divide classes unevenly. He compares them to cranes who suppose that the world is divided into cranes and all other animals, or to jingoistic Greeks who see the world as Greeks and barbarians. It is hard to escape the sense that Socrates is making just this mistake in this passage of the *Republic*.

Again, there is something wrong with Socrates' use of the words 'being', 'are' and 'really' throughout this passage in the *Republic* and all over his work. He seems to think, as we would say, that ideas are 'more real' than physical things. This is implicit in his distinction between 'being' and 'becoming' as if 'becoming' was somehow less real than 'being'. But in the Greek language as in English there is no sense to be given to 'more real', 'most real'. It is not a comparative or superlative adjective. The use of these terms in the way Socrates uses them is an abuse, and it is one that his teacher Parmenides never made. There are no distinctions of higher or lower in the Parmenidean order, but everything is equally full of being.

To the detailed objections here the answer is the same in both cases. The reason why the unity in the created order is less than that in the intelligible realm is the reason why *eikasia* is less aware than *dianoia* that visible objects image ideas. In both cases we are comparing modes of opinion with modes of knowledge. The unity to which the Demiurge looks is not the Good but the intelligible, living creature which images the Good. The prisoners in the Cave, like the geometers, have no real knowledge, but they are as much deeper in ignorance as opinion is darker than knowledge. And so they believe that what they see is all there is. In this way these differences between *pistis* and *noesis*, between *eikasia* and *dianoia* actually confirm the general pattern which Plato has imposed on them and the gradations in ignorance follow through as they should.

As for the more general objections, the first was that Socrates has made three categories out of tiny minorities and the fourth out of all the rest, while in the *Statesman* Plato speaks against such divisions. But just as *noesis* is the very highest level of human understanding, so *eikasia* is the very lowest. To suppose that this material world around us and our own bodies constitute the sum of things is to be as ignorant as it is possible for a human being to be. In this way the Line has as its terms the two limits of human cognition. And though it is true that Plato speaks in the *Statesman* as he does, it is also true that the greatest division of all between the members of humankind is between those who are realized and those who are not. So Homer suggests that there were Northern gates through which humanity at large is born and Southern gates through which Odysseus alone passes as an immortal.[1] Heraclitus believed that one man was worth ten thousand if he was the best. And Parmenides is told by the Goddess who is waiting for him and him alone, that his journey to her is far from the path of men. This is elitist to a degree, and certainly fits much else in the *Republic*.

The account we have given of *pistis* allows us to place Plato's own understanding of natural science on the Line and this is surely necessary if the scheme is intended to be exhaustive. Even so the fourth category of *eikasia* seems enormous. It must include not only the bare belief that material phenomena alone are real, but all those skills founded on experience which Plato describes as empirical.

---

1. *Odyssey xiii*, 110–111.

Cooking is his chief example. The capacity to manipulate materials to given ends simply on the basis of memory without any theory or explanation of how those causes achieve these effects is the hallmark of the debased understanding. Compare the drugs we use for mental problems. According to Leibniz this kind of understanding constitutes the entirety of some people's grip on things. And Leibniz supposed that this kind of thinking was within the reach of animals too, so that such thinking provides a bridge between the human and animal realms.[2] Plato believed that the souls of humans and animals transmigrated into animals and humans, and that animals were less capable than humans of contemplation because the roundness of their skulls had been permanently deformed.

So much may be said in defence of Plato. On the other hand the charge that in this passage Socrates misconstrues the grammar and logic of the verb 'to be' and its parts is, I think, impossible to defend. Like 'dead' and 'unique', 'real' cannot be used in the comparative or superlative. Being does not have levels, and becoming is as fully, is as real as anything can be. Plato gets this wrong throughout this passage. In the preceding book of the *Republic* Socrates makes an even worse blunder over the logic of the verb 'to be'. He supposes that because any beautiful particular will sometimes appear ugly, it 'rolls around' between being and non-being. This is a clear confusion of the predicative with the existential use of the verb 'to be'. That something is no longer beautiful does not mean that it has ceased to exist. If it has become ugly, it now exists as ugly.

How important is this to Socrates' account of the modes of cognition? Is it possible to sustain the body of his account after removing these errors? I think so, but without this rhetoric it would somehow no longer be Plato. The fundamental element here is the threefold repetition of the relationship or ratio of imitation. Provided that is maintained, the theory still stands. There can be no question that the world of becoming is equally real. There can be no admixture or dilution of being by some 'non-being' which thins it out and makes it temporal.

In another way, the scheme of the Divided Line and the Cave

2. *New Essays on Human Understanding*, p. 143.

does not break Parmenides' great law that everything is full of being. The prisoners in the Cave are as ignorant as it is possible to be but the spectacle which they observe on the back wall is projected by the puppets and the fire, and they in turn image natural objects and artefacts outside the Cave. The landscape through which the freed prisoner travels to the final vision of the sun is the intellectual landscape behind every single prisoner. With the machinery of the Cave, that landscape forever generates the shadowplay on the wall, whether the prisoners realize it or not. At their core they are in eternity; the Good radiates from within them no less and no more than it does in the realized philosopher. It must do or they could not be experiencing what they do experience.

We began with the mathematical idea of the Equals themselves. We noted then that this idea was markedly different from our abstract notion of equality. Plato's idea of equality is very different from the equality which things have to other things in the world, but it is only by our prenatal knowledge of the Platonic idea of the Equals that we can ever recognize those equal things in the world. Furthermore equal things in the world are equal to each other only by virtue of their participating in or imitating the Equals themselves. Put in these terms Plato's theory of ideas is problematic.

In the *Phaedo* Socrates describes how as a young man he became frustrated by the apparent contradictions in so-called scientific thinking. How could a unity become a duality? Either by adding another or by dividing it into two. But then quite different processes, addition and division, produce the same result. To avoid these difficulties Socrates resorted to a simple verbal stratagem. When asked how it was that one thing could become two, he would reply, 'By twoness'. How did things become big? 'By bigness'.

So far so good. But Socrates was also an ethicist. Here too the question 'What makes someone courageous?' may be answered truthfully and vacuously by 'Courage'. But in this case Socrates really did believe that there was some spirit or energy in the courageous person making them so. He thought that this inward power was a kind of knowledge, knowing what was worth dying for, and the person who had this knowledge as knowledge acted courageously. So this knowledge, courage, actually propelled the courageous into their courageous deeds. In Socrates' view much the same could be said of all the virtues. They were all active energies propelling those who had certain kinds of knowledge into virtuous action. For Socrates the dialectical problem then became how to distinguish between the various different virtues, since they were now all forms of knowledge.

And so the claim that things in the world are equal by means of the Equals themselves became more than a verbal stratagem to avoid the paradoxical reasoning of the sciences. The Equals, bigness, unity became actual agents in the shaping of the world and our thinking about it. If the ideas have such powers, they must themselves have the properties with which they invest particulars in the world. The idea of the Beautiful must be quintessentially beautiful, whose radiance emanates so as to inspire all that is beautiful around us.

Insofar as the theory of ideas is used to explain how we come to see things as equal, for example, this is not a problem. By virtue of our prenatal knowledge of the idea of the Equals we see things here as equal, and we could not see things here as equal if we did not have that foreknowledge of the Equals. But the invisible Equals, the idea of Justice, the beautiful itself by itself are not merely principles of the understanding. They are the actual causes and makers of their instantiations on earth in time and space. This ontological claim is much more difficult to make out.

There are two ways of doing it. They are mutually incompatible and Plato adopts both. The easier way is what he does in the *Timaeus*. He supposes a Craftsman who looks to the ideas and especially the perfect, intelligible living creature. The craftsman does not make this paradigm. It is given and he realizes it in part in his making of the stellar and planetary systems. This is easy for us to understand for many reasons. It follows Homer in his account of how Hephaestus made the shield of Achilles and the palaces of the Gods. It is closely comparable, as Philojudaeus showed, to the very first chapters of the Hebrew *Genesis*, and so is particularly familiar to

Christian readers. And the notion that the carpenter or smith looks to the idea of the thing to be made, and must be contemplative as well as practical, is the basis of Plato's theory of the arts and crafts.

But philosophically the importation of the Craftsman to explain how invisible ideas shape the physical world, does no good at all. Now in addition to the invisible ideas we have an invisible Craftsman as well, and we are no further in our search but have even more to explain. This easier solution turns out to make the problem even worse. The harder solution is the one adopted in the *Republic*. In fact it is no solution but at least it does not exacerbate the problem. Here there is no Craftsman to convert the ideas physically into our world. The ideas do it by themselves. The idea of the Good generates the sun and the sun makes everything here and makes it visible to us. As Plato says of the noetic mathematician who achieves the vision of the Good, that journey starts, ends and never for a moment deals with anything but ideas. No gods here.

The *Republic* is emanationist where the *Timaeus* is creationist. And this notion that the physical creation is spontaneously generated by the invisible ideas alone is taken up by Plotinus who has Nature say:

> The mathematicians from their vision draw their figures: but I draw nothing: I gaze and the figures of the material world take being as if they fell from my contemplation.[3]

Nature explains how she herself is the precipitate of yet higher beings, in contemplation. There is no designing, no forethought, no deliberate organization. All the orders of beings spring from the One in due succession effortlessly. All that the higher powers have to do is to fix their minds on the order of being above them and that contemplation by itself is enough to generate the physical world as we have it. This account by Plotinus does not explain how invisible ideas shape the physical world, but it gives us a picture of the process which is more persuasive than the picture of the Craftsman in the *Timaeus*. Plotinus' account is much closer to Socrates' description of Sun and Good.

3. *Ennead* III, 8.

These two ways of understanding how the ideas organize the creation, with and without a Demiurge, are mutually incompatible. But they are convergent. Plotinian Nature contemplates the All-Soul who contemplates the Reason-principles who contemplate the One. There are dimly recognisable minds at work here, for all that they do not consciously create. On the other hand the Demiurge does not create the ideas and paradigms after which he makes the cosmos. And sometimes the ideas are more than just models and are the substantial material of the creation. Such is Timaeus' claim about the Same, the Different and Being which are mixed together to make the soul-stuff. It is from this soul-stuff that the Demiurge makes the circuits of the Same and the Different, after dividing it in the portions of the *lambdoma,* the means and the tones. From a certain point of view the description of the mixing of these three ideas is unusually primitive even for Plato.

Sameness and difference as a pair are also mentioned in another context in the *Sophist.* Here they are added to three 'greatest' or 'highest' kinds: being; motion; rest. Plato then explores the logical or dialectical relations between these five terms in order to show that a certain sense can be given to the expression 'non-being', despite Parmenides' claim that 'non-being' was unspeakable and unthinkable. Since motion, say, is different from being though a part of being, it is non-being insofar as it is different from being, but it is still part of being and so also being. In this limited sense 'non-being' is possible.

Could the dialectical relations between 'the three greatest kinds' and sameness and difference somehow create the circuits of the Same and the Different? The dialectical relations between sameness and difference are very complex. As opposites they are of equal power, and the circuits of the Same and the Different are quantitatively identical. Likewise Sameness and Difference are equally mixed with being to make the soul-stuff from which the circuits are made. The domination of Difference by Sameness in other respects may reflect the logical priority of sameness over difference through the kind of analysis we find in the latter part of the *Parmenides.* As for motion, it belongs here since it is the defining character of soul, which is what moves itself. The place of rest in the psychic and physical complex is less clear. Everything seems to be in motion, and even the still earth in the centre is claimed once by Plato to 'weave' around its axis. But the essential rest of the cosmos, in its relation to motion, is one of the most beautiful features of the system. For all of it turns, geocentrically, within its own space:

> Remaining the same in the same place it rests by itself and so remains fixed.[4]

Do the circuits of the Same and the Different express the dialectical relations between Sameness and Difference themselves, and then between these two and Being, Motion and Rest? If so, our mathematical analysis of Timaeus' account is at best only half the story, and a dialectical account is needed alongside it. But that is beyond the scope of this book. How do invisible ideas shape the world ontologically as well as epistemologically? We must imagine the ideas not as passive and static but as active, overflowing with energy in their interrelations with cognate ideas. The greatest ideas, the most general, have the most complex interrelations which are at once eternally fixed and in their roiling entanglement the very energies which drive the stars and planets. For it is not finally love which moves the Sun and all the stars but that glorious excess of the ideas in eternity. And since Motion and Rest are as much a part of music as they are of astronomy, along with Being, Sameness and Difference, music too may be their overflow. This is what is adumbrated by the Pythagorean oath.

4. Parmenides, 8, 29–30.

12

Music

THE UNREASONABLE effectiveness of mathematics in the natural sciences is an issue for philosophers of science. In the third chapter I have suggested that both Kant and Plato anticipated this problem and solved it before anyone else had propounded it. But there is another problem which arises from the relations between mathematics and the physical world. And that is the unreasonable effectiveness of music over the natural affections. Music, as Leibniz said, is the activity of a mind counting while unconscious that it does so.[1] Why does this unconscious counting so affect our feelings? That the mathematical realm should precede the physical realm and inform it rests on the identity of the two worlds. The major difference between them for Kant was that the physical realm is one spatio-temporal whole while mathematical space and time are not. But there is no such obvious identity between the mathematical intervals and the stirrings of hearts on our hearing music. That mathematics should fit the world of physical forms and forces is much less surprising than that it should control the inner world of our passions.

This, presumably, is the reason for the forging of the circuits of the stars from bands made of the musical intervals. First the Demiurge makes the long strip by placing all the quantities of soul-stuff in incremental order starting from one and ending with twenty seven. Then he halves and quarters this strip to make the orbits of the Equator and the planets. Halved or quartered the strip is composed of the same intervals in the same order. When the second quarter-strip is divided into the six inner orbits, only then do we

1. *Principles of Nature and of Grace,* 17.

move from the larger orbits which comprise all the intervals, to smaller orbits of which each has different portions of the nearly six octaves. But integral or partial, the orbits of the world-soul are composed of this long scale, and this must help to explain why music has such power over the soul. There is a sympathetic vibration from an organism very similar in its structure and components to the musical order. For not only are the orbits composed of these mathematical and musical intervals. The orbits are then arranged in accordance with the same order in their relations to each other.

A human soul, so composed and so arranged, responds to musical sounds. There is nothing remarkable about this, since human souls respond to all their sensations, which enter through their sense-organs, often to overwhelming effect. But Plato supposes that rational musical sounds have a very rare power indeed, and that is to restore the overwhelmed soul to its proper order again. The only other activity to have this same restorative power is the study of motion in astronomy. Generally, sensory experience is disruptive of the soul's circuits, especially in people still growing up. Then the circuit of the Same is completely stopped, which means in effect that the intellect's perception of the eternal is entirely obscured. And the circuits of the planets are twisted out of all recognition by the violence of physical sensation in the early years. But musical sounds can restore the proper order for a while even in these years and return the soul to its primeval calm and cheerfulness. After all, Pythagoras had once quietened a lunatic by playing on his lyre.

The same skill which enabled a musician to quieten an audience could also be used to stir and agitate it, to invade the harmonious revolutions of the soul. This was a power of which Plato was acutely conscious as a lyric poet himself. He was a musical conservative and describes in the *Laws* the superiority of the Egyptians over all other peoples at least in this, that they had vigorously resisted all impulses to musical innovation but had strictly and exclusively preserved the music of their earliest ancestors.[2] He explains how the young of all creatures are by nature restless, so children should be trained to sing and dance well. He seems to think that the songs and dances of

2. *Laws*, 656d–657b.

childhood are invitations to the children to join in the festivities of the Gods, as though they dance and sing with stars.

In the *Republic* Socrates approves certain kinds of music as suitable for his citizens, the Dorian and Phrygian styles. For the other styles his citizens would have no use. Socrates wants a music which strengthens people in adversity and another to calm them in meditation.[3] We may compare and contrast this with the key signatures of Classical music which were also associated with dominant feelings. But nowhere in Plato's work do we find an even partial answer to the question with which we began: how do musical sounds influence human feelings? We come much nearer to answers with the musical doctrine of affections in the eighteenth century than we do in Plato's work.

Plato's theory of the human soul as an astronomical microcosm has its own beauty. The theory was profoundly shaped by his susceptibility to music. His account of the psychic disruption occasioned by the senses as we grow turns on the damage this does to the judgement. A soul so confused is incapable of discriminating accurately between the Same and the Different as it encounters the things of sense. But the restoration of the soul from that disorder is particularly easy to understand as the experience of listening to music after a troubled time. Plato describes how the stopped circuit of the Same and the twisted circuits of the Different unravel again. And this is not a fortuitous and incidental feature of his account of the soul but crucial to it. He was describing what music does to us actually in a theory all too theoretical. I cannot detect the circuits within my soul either when they are disorganized or when they are harmonious. But I can detect them in that moment of transition when I pass from cares to musical attention.

I have claimed in Chapter Seven that the planets and stars do not make sounds in their courses. These courses are related to each other as are the numbers of the *lambdoma* to each other. In this way

3. *Republic*, 399b.

the visible heavens manifest the same order as the nearly six octaves of the diatonic scale. It is the correlation between the visible and audible realms which Timaeus promotes, not the notion of a sounding universe. But if this is true of the *Timaeus,* it is not true of the *Republic* where the rim of each bowl has a Siren seated on it who utters one note as she is carried round. Of course, it could be argued that the singing Siren was merely a way of pointing up the identity between the heavenly courses and the diatonic scale, but the issue is deeper than this even in the *Timaeus.*

When we think of the seven numbers of the *lambdoma,* we naturally think of them as seven cardinal numbers, seven numbers in the series of twenty seven numbers which stretches from 1 to 27. But this is wrong. As presented in the *Timaeus* the seven numbers of the *lambdoma* belong only in the series of thirty four numbers which stretches from the first note in the first octave to the sixth note in the fifth octave. This is the series to which the lengths of the planetary courses belong. In order then to understand their interrelations we can do no better than to imagine them as musical intervals. Each planet is related to the other six by six different intervals. But we must not forget that it is the length of the planetary circuit, not the planet, which is being harmonized in this way. Unlike the moving planet, its circular course is fixed and constant.

But if we allow ourselves to localize the planet as opposed to its course, as Socrates places his Sirens at one point only on each of the rims, then we have the makings of a musical astrology. We might take the order in which the seven planets appear above the horizon in the course of the day to construct a melody with appropriate pauses. Or we might construct our melody from the closeness of the planets to us latitudinally, on the basis that it is whether they are summering or wintering that most affects their influence. And the horoscope of each of us could be expressed in a great chord of seven notes in which the dynamic of each note would be determined by the planet's proximity to the place of birth. But, of course, any authentic Platonic astrology begins from the planets' equatorial moments in the *Phaedrus.*

In these ways and to this extent there is a music of the spheres in Plato. We may imagine the planets as tones and their interrelations as the intervals between the tones. This is as immediate a way of understanding the heavens as looking at them, and much more easily reproduced. And it raises that difficult question of the relative priority between the cardinal and the musical series of numbers. How can we compare the quality and clarity of the cardinal numbers against the musical numbers? Do I apprehend arithmetic with the same lively assurance I hear music? How can I tell? Again, it is easy enough to represent the musical series in terms of the cardinal numbers. It is much more difficult to represent the cardinal numbers in terms of the musical tones.

The musical series is geometric and its terms are proportionate to each other. It is always a matter of ratios. The arithmetical series is also cumulative but always by the same quantity. Here it is a matter of strictly measuring quantities by an agreed unit. Where the musical or geometric series is central, the culture is musical, conscious of appearances and much given to metaphor. Where the arithmetical series dominates, the culture is unmusical, utilitarian and sceptical.

13

Science as Worship

Eratosthenes, in the book entitled *The Platonist,* reports that the Delians, having consulted the oracle as to how to save themselves from the plague, were prescribed by the God to construct an altar double the size of the one which already existed. This problem threw the architects into a strange embarrassment. They wondered how one could make one solid the double of another. They questioned Plato about this difficulty. He answered that the God had thus sent this divination not because he had any need of a double altar, but in order to reproach the Greeks for neglecting the study of mathematics and for belittling the value of geometry.[1]

THIS IS one of the most outrageous Delphic oracles of which we hear. Delos was an island sacred to Apollo and the largest slave-market in the Aegean. Plague there could have disastrous consequences. And the anxious Delians must solve a problem in solid geometry! The gulf between the crisis and the advice is cruel. How similar is the situation here to the one with which the *Iliad* opens? Apollo afflicts the Greek army with plague because Agamemnon abused his priest, Chryses. Chryses prayed to Apollo Smintheus, the Mouse God, and Apollo sends the plague. That plague was withdrawn by Apollo when Agamemnon yielded to Chryses and Apollo was given several hecatombs. Did Apollo likewise cause the Delian plague to spur the Greeks to geometry? We are certainly not told so, but why should doubling his altar end it?

Apollo's demand of the Delians is much more troublesome than

1. *Theon of Smyrna,* trans. Lawlor (San Diego, CA: Wizards Bookshelf, 1979), pp. 1–2.

it looks. The altar was a cube and the doubling of a cube is not like the doubling of a square which Socrates demonstrates in the *Meno*. That is easily done by constructing the square on a diagonal of the given square. But a cube is not to be doubled by any of the geometrical methods known to the Greeks. So Apollo's choice of problem was acute. It raises the difficulties of dealing with incommensurables in a new way, which could not be handled by resorting to plane geometry. That shift accommodated such incommensurables as the side to the hypotenuse of an isosceles right-angled triangle, but there was no clean solution at all to the problem which Apollo had set. And this was how Plato read it. The God had set what seemed to be an insuperable problem in order to drive the Greeks to greater effort in mathematics. Unfortunately our version of the story does not tell us what happened with the plague.

Clearly the Delians regarded Plato as a living authority on mathematics, and he for one was certainly spurred into action by the oracle. Unable to find a solution by a mathematics of proportions, Plato developed a mechanical instrument to enable the determining of the side of a doubled cube. And we hear an echo or a premonition of all this, perhaps, in the setting out of the mathematical sciences in the *Republic*.[2] Socrates passes from plane geometry to astronomy as the study of solid bodies in motion. But he catches himself and points out that they have missed a step, since solid geometry comes after plane geometry but before solid bodies in motion. Glaucon here as elsewhere seems to Socrates a little less enthusiastic for all these 'useless' sciences than he should be, and Socrates expounds on how much some state support would do for the proper study of cubes and depth. We do not know how generously the Delian treasury responded to their crisis and the Oracle's reply.

Apollo, according to Plato, wanted to encourage the study of mathematics. He chose a grim occasion to make his want known. And there is an indication of Apollo's concerns in this matter in the *Republic*. As given, the nuptial number is a jest of the Muses and a rather heavy-handed one. But I have explained how this number

2. *Republic*, 528b–e.

and the figure which it generates are the first mathematical demonstration of the distance between the Tropics. In the history of scientific geography this is a most significant result. And here it is attributed to the Muses, goddesses of Parnassus, attendants on Delphic Apollo, in full mathematical flight. Of course the Muses may be found also on Mount Helicon and elsewhere, but mathematical Muses are likely to be Delphic. Like the oracle about the doubling of the cube the nuptial number is presented as a puzzle. Plato's recommendation that astronomy and music, in particular, be studied by the setting and solving of puzzles is also Delphic practice.

For his part Timaeus gives an extended account of the five regular solids, and looks rather to planetary astronomy for the research of the future. Not that this research would be entirely original, since, few though they are, some people have yet managed to discover these things in earlier times. In the *Epinomis* the emphasis on astronomy is maintained, and the sense of its enormous difficulty. Greek astronomy is placed in its relations to the wisdom of Syria and Egypt whose climate permits a much broader vision of the night sky. But the Athenian stranger in the *Epinomis* has a specific vision of scientific progress for the Greeks. Through the study of astronomy, the Greeks will yet outstrip their predecessors and contemporaries in star wisdom, because the Greeks have always improved upon their acquisitions from abroad.

This scientific program is a most pious form of worship. The stars are either Gods themselves or the most beautiful handiwork of Gods. Either way, studying the stars is a way to knowing the divine. Though the Greek climate is inferior to the Syrian and Egyptian for star-gazing, the Greeks have some great advantages. They have the teaching and the oracles of the Delphic oracle, as well as the religious ritual enshrined in their laws. The *Epinomis* is clear that Apollo at Delphi actively promoted these studies as a means to spiritual development. And the dangers of theocratic nationalism are balanced by the Athenian stranger's recognition that other nations are far ahead of the Greeks in just this knowledge.

The scientific program is a powerful form of worship because the gods are very well aware of our practising mathematics:

.... the Divine is never either unintelligent or in any ignorance of human nature, but knows that if it teaches us we shall follow its guidance and learn what is taught us. That it so teaches us, and that we learn number and numeration, it knows, of course ... wholeheartedly rejoicing with one who has become good by God's help.[3]

3. *Epinomis*, 988b, trans. A. E. Taylor.

Made in United States
North Haven, CT
09 October 2022

25215176R00082